D0936113

A Guide to Elementary Number Theory

Library of Congress Catalog Card Number 2009937592
ISBN 978-0-88385-347-4
Printed in the United States of America
Current Printing (last digit):
10 9 8 7 6 5 4 3 2 1

The Dolciani Mathematical Expositions
NUMBER FORTY-ONE

MAA Guides # 5

A GUIDE TO ELEMENTARY NUMBER THEORY

Underwood Dudley

Published and Distributed by
The Mathematical Association of America

The DOLCIANI MATHEMATICAL EXPOSITIONS series of the Mathematical Association of America was established through a generous gift to the Association from Mary P. Dolciani, Professor of Mathematics at Hunter College of the City University of New York. In making the gift, Professor Dolciani, herself an exceptionally talented and successful expositor of mathematics, had the purpose of furthering the ideal of excellence in mathematical exposition.

The Association, for its part, was delighted to accept the gracious gesture initiating the revolving fund for this series from one who has served the Association with distinction, both as a member of the Committee on Publications and as a member of the Board of Governors. It was with genuine pleasure that the Board chose to name the series in her honor.

The books in the series are selected for their lucid expository style and stimulating mathematical content. Typically, they contain an ample supply of exercises, many with accompanying solutions. They are intended to be sufficiently elementary for the undergraduate and even the mathematically inclined high-school student to understand and enjoy, but also to be interesting and sometimes challenging to the more advanced mathematician.

1. *Mathematical Gems,* Ross Honsberger
2. *Mathematical Gems II,* Ross Honsberger
3. *Mathematical Morsels,* Ross Honsberger
4. *Mathematical Plums,* Ross Honsberger (ed.)
5. *Great Moments in Mathematics (Before 1650),* Howard Eves
6. *Maxima and Minima without Calculus,* Ivan Niven
7. *Great Moments in Mathematics (After 1650),* Howard Eves
8. *Map Coloring, Polyhedra, and the Four-Color Problem,* David Barnette
9. *Mathematical Gems III,* Ross Honsberger
10. *More Mathematical Morsels,* Ross Honsberger
11. *Old and New Unsolved Problems in Plane Geometry and Number Theory,* Victor Klee and Stan Wagon
12. *Problems for Mathematicians, Young and Old,* Paul R. Halmos
13. *Excursions in Calculus: An Interplay of the Continuous and the Discrete,* Robert M. Young
14. *The Wohascum County Problem Book,* George T. Gilbert, Mark Krusemeyer, and Loren C. Larson
15. *Lion Hunting and Other Mathematical Pursuits: A Collection of Mathematics, Verse, and Stories by Ralph P. Boas, Jr.,* edited by Gerald L. Alexanderson and Dale H. Mugler
16. *Linear Algebra Problem Book,* Paul R. Halmos
17. *From Erdős to Kiev: Problems of Olympiad Caliber,* Ross Honsberger

18. *Which Way Did the Bicycle Go? ... and Other Intriguing Mathematical Mysteries,* Joseph D. E. Konhauser, Dan Velleman, and Stan Wagon
19. *In Pólya's Footsteps: Miscellaneous Problems and Essays,* Ross Honsberger
20. *Diophantus and Diophantine Equations,* I. G. Bashmakova (Updated by Joseph Silverman and translated by Abe Shenitzer)
21. *Logic as Algebra,* Paul Halmos and Steven Givant
22. *Euler: The Master of Us All,* William Dunham
23. *The Beginnings and Evolution of Algebra,* I. G. Bashmakova and G. S. Smirnova (Translated by Abe Shenitzer)
24. *Mathematical Chestnuts from Around the World,* Ross Honsberger
25. *Counting on Frameworks: Mathematics to Aid the Design of Rigid Structures,* Jack E. Graver
26. *Mathematical Diamonds,* Ross Honsberger
27. *Proofs that Really Count: The Art of Combinatorial Proof,* Arthur T. Benjamin and Jennifer J. Quinn
28. *Mathematical Delights,* Ross Honsberger
29. *Conics,* Keith Kendig
30. *Hesiod's Anvil: falling and spinning through heaven and earth,* Andrew J. Simoson
31. *A Garden of Integrals,* Frank E. Burk
32. *A Guide to Complex Variables* (MAA Guides #1), Steven G. Krantz
33. *Sink or Float? Thought Problems in Math and Physics*, Keith Kendig
34. *Biscuits of Number Theory*, Arthur T. Benjamin and Ezra Brown
35. *Uncommon Mathematical Excursions: Polynomia and Related Realms*, Dan Kalman
36. *When Less is More: Visualizing Basic Inequalities*, Claudi Alsina and Roger B. Nelsen
37. *A Guide to Advanced Real Analysis* (MAA Guides #2), Gerald B. Folland
38. *A Guide to Real Variables* (MAA Guides #3), Steven G. Krantz
39. *Voltaire's Riddle: Micromégas and the measure of all things*, Andrew J. Simoson
40. *A Guide to Topology*, (MAA Guides #4), Steven G. Krantz
41. *A Guide to Elementary Number Theory*, (MAA Guides #5), Underwood Dudley

MAA Service Center
P.O. Box 91112
Washington, DC 20090-1112
1-800-331-1MAA FAX: 1-301-206-9789

Introduction

This *Guide* is short, so this introduction will be short as well.

It contains most of the topics that are typically part of a first course in elementary number theory as well as some others that could be. It is intended for two kinds of readers. The first are those who once knew but have forgotten, for instance, which integers are the sum of two squares and why they have that property, and who want their memories refreshed. The second are those who were never told about sums of two squares and want to find out about them, quickly and with no nonsense. The *Guide* is not a textbook—for one thing there are no exercises—but I like to think that someone new to elementary number theory could go through it and come out ahead.

The *Guide* does not contain some things that I will mention here lest readers think that they were omitted through carelessness. There is no material on cryptography because that is an application of number theory. Elliptical curves are intermediate, not elementary, number theory. Only a few well-known diophantine equations appear. Quadratic forms are missing, in spite of the hundreds of hours that Gauss so profitably devoted to them. Algebraic number theory is a whole other subject. Finally, there are no references to the literature or suggestions for further reading. In these electronic days, the Web has a vast amount of material on any subject in the *Guide*, growing vaster and vaster every day, that readers can find with a few clicks of the mouse. However, nowhere in cyberspace can all of its contents be found in as compact a form.

I hope it is of use.

Thanks are due to those who read the manuscript closely. Their comments reduced the number of errors it contained and increased its clarity. That the first is not now zero and the second less than infinite is my fault.

Underwood Dudley

Contents

CHAPTER 1

GREATEST COMMON DIVISORS

The main concern of elementary number theory is the integers, $\ldots, -2, -1, 0, 1, 2\ldots$, especially those that are positive.

We say that d divides a if a/d is an integer:

Definition $d \mid a$ ("d divides a") if there is an integer k such that $a = kd$.

Thus $3 \mid 24$ and $5 \mid -100$. Zero does not divide any non-zero integer.

If d divides each of a set of integers, it divides any linear combination of them. Here is a proof for a set of size two—the general case is the same except for more complicated notation.

Theorem *If $d \mid a$ and $d \mid b$ then $d \mid (ar + bs)$ for any integers r and s.*

Proof We know that there are integers k and j such that $a = kd$ and $b = jd$. Then

$$ar + bs = kdr + jds = (kr + js)d.$$

Thus $d \mid (ar + bs)$.

So we know that $6x + 21y$ is a multiple of 3 for any integers x and y.

Two integers always have at least one positive divisor in common, namely 1, and may have more. We have a notation for the greatest of them:

Definition The *greatest common divisor* of non-zero integers a and b, denoted (a, b), is the integer d with the properties that $d \mid a$ and $d \mid b$ and that if $c \mid a$ and $c \mid b$, then $d \geq c$.

For example, $(6,21) = 3$ and $(5,21) = 1$.

We give a name to integers with no positive common divisor other than 1:

Definition If $(a,b) = 1$, then we will say that a and b are *relatively prime*.

There follows a theorem whose truth is obvious, but whose proof we include to insure that our intuition has not led us astray. It will be used often later.

Theorem If $(a,b) = d$ then $(a/d, b/d) = 1$.

Proof Let $(a/d, b/d) = c$. We know that $c \mid a/d$ and $c \mid b/d$ so there are integers r and s such that $a/d = cr$ and $b/d = cs$. That is, $a = (cd)r$ and $b = (cd)s$ so cd is a common divisor of a and b. Common divisors can be no larger than greatest common divisors, so $cd \leq d$. Thus $c \leq 1$. But c is a greatest common divisor, so $c \geq 1$. We conclude that $c = 1$.

It is possible to take as an assumption the property that dividing one integer by another gives an integer quotient and a remainder that is less than the divisor. Rather than do that, we go further back and assume that a non-empty set of nonnegative integers has a smallest element.

Theorem (the division algorithm) If a and b are positive integers then there are unique integers q and r such that

$$a = bq + r, \quad 0 \leq r < b.$$

Proof The subset consisting of the nonnegative integers in

$$\{a, a-b, a-2b, a-3b, \ldots\}$$

is nonempty because a is positive, and so it contains a smallest element, call it $a - qb = r$. We know that $r \geq 0$ because r is nonnegative and that $r < b$ because if $r \geq b$ then $r - b = a - (q+1)r$ would be nonnegative, contradicting the fact that $a - qb$ is the smallest nonnegative element of the set.

The construction shows that q and r are unique, but if that is insufficiently convincing here is a proof. Suppose that $a = bq + r = bu + v$ with r and v in the interval $[0, b-1]$. Then $0 = b(q-u) + (r-v)$ so $b \mid (r-v)$. But $-(b-1) \leq r - v \leq b - 1$. The only multiple of b in that interval is 0, so $r = v$ and $0 = b(q-u)$. Thus $q = v$ and so the quotient and remainder are unique.

For example, dividing 27 by 10 gives a unique quotient of 2 and remainder 7 : $27 = 2 \cdot 10 + 7$.

The division algorithm, together with the next lemma, will give us the Euclidean algorithm, a method for finding greatest common divisors by repeated subtraction.

Lemma If $a = bq + r$, then $(a, b) = (b, r)$.

Proof Let $d = (a, b)$. Then $d \mid a$ and $d \mid b$ and from $a = bq + r$ it follows that $d \mid r$. Thus d is a common divisor of b and r. Let $c = (b, r)$, so $c \geq d$. Because $c \mid b$ and $c \mid r$ from $a = bq + r$ it follows that $c \mid a$. Thus c is a common divisor of a and b. Thus $c \leq d$. That, together with $c \geq d$, shows that $c = d$.

Theorem (the Euclidean algorithm) *If a and b are positive integers and*

$$\begin{aligned} a &= bq + r, \quad 0 \leq r < b, \\ b &= rq_1 + r_1, \quad 0 \leq r_1 < r, \\ r &= r_1 q_2 + r_2, \quad 0 \leq r_2 < r_1, \\ &\vdots \\ r_k &= r_{k+1} q_{k+2} + r_{k+2}, \quad 0 \leq r_{k+2} < r_{k+1}, \end{aligned}$$

then for k large enough, say $k = t$, we will have $r_t = r_{t+1} q_{t+2}$ and $(a, b) = r_{t+1}$.

Proof The decreasing sequence of nonnegative integers $b > r > r_1 > r_2 > \cdots$ must end with 0. Suppose $r_{t+2} = 0$, so $r_t = r_{t+1} q_{t+2}$. Applying the last lemma repeatedly

$$(a, b) = (b, r) = (r, r_1) = (r_1, r_2) = \cdots = (r_t, r_{t+1}) = r_{t+1}.$$

An important consequence of the theorem is the

Theorem *If $(a, b) = d$ then there are integers x and y such that $ax + by = d$.*

Proof The idea of the proof is to take $r_{t-1} = r_t q_{t+1} + r_{t+1}$ to get $(a, b) = r_{t+1}$ in terms of r_t and r_{t-1} and then back-substitute until it is in terms of a and b. To avoid burdensome notation, we will give an example that could if needed be converted into a general proof.

The forward calculation of $(1986, 1494) = 6$ is

$$\begin{aligned}1986 &= 1 \cdot 1494 + 492\\ 1494 &= 3 \cdot 492 + 18\\ 492 &= 27 \cdot 18 + 6\\ 18 &= 3 \cdot 6.\end{aligned}$$

The back-substitution is

$$\begin{aligned}6 &= 492 - 27 \cdot 18\\ &= 492 - 27(1494 - 3 \cdot 492)\\ &= 82 \cdot 492 - 27 \cdot 1494\\ &= 82(1986 - 1 \cdot 1494) - 27 \cdot 1494\\ &= 82 \cdot 1986 - 109 \cdot 1494.\end{aligned}$$

So $(1986, 1494) = 1986x + 1494y$ with $x = 82$ and $y = -109$.

Here is an alternative proof, not by example. If a and b are positive, the set of linear combinations of a and b contains positive elements, including $a \cdot 1 + b \cdot 0$ and $b \cdot 0 + a \cdot 1$, and so has a smallest positive element, $t = ax + by$. The division algorithm says that $a = tq + r$ with $0 \le r < t$. Then

$$r = a - tq = a - (ax + by)q = a(1 - xq) + b(-yq).$$

So r is a linear combination of a and b. Because t was the smallest positive linear combination of a and b, r must be 0. Thus $a = tq$ and $t \mid a$. Similarly, by dividing b by t, we can show that $t \mid b$. Thus t is a common divisor of a and b, so $t \le (a, b)$. Because $t = ax + by$ and (a, b) divides both terms on the right, $(a, b) \mid t$ and hence $t \ge (a, b)$. Thus $t = (a, b)$.

The theorem has an important corollary.

Corollary *If $d \mid ab$ and $(d, a) = 1$, then $d \mid b$.*

Proof From the theorem we know that there are integers x and y such that $dx + ay = 1$. Thus $d(bx) + (ab)y = b$. Because $d \mid ab$, both terms on the left are multiples of d, and hence so is their sum. That is, $d \mid b$.

The condition that $(d, a) = 1$ is necessary. For example, $6 \mid 8 \cdot 9$ but 6 divides neither of the factors.

Another corollary, obvious once we have the Unique Factorization Theorem (which we will not until the end of the next chapter) can be proved without it:

Corollary *If $a \mid m, b \mid m$, and $(a, b) = 1$, then $ab \mid m$.*

Proof Because $b \mid m$ we know that $m = bq$ for some q. Because $a \mid m$ we have $a \mid bq$. But $(a, b) = 1$ so from the first corollary we know that $a \mid q$. That is, $q = ar$ for some r. Thus $m = bq = bar$. This shows that $ab \mid m$.

The necessity of $(a, b) = 1$ is shown by the example $3 \mid 24, 6 \mid 24$, $18 \nmid 24$.

A companion to the greatest common divisor is the least common multiple.

Definition If m is a positive integer multiple of a and of b, and if every other positive common multiple of a and b is larger than m, then $m = [a, b]$ is the *least common multiple* of a and b.

For example, $[10, 6] = 30$.

After we have the Unique Factorization Theorem, that each integer has a unique representation as a product of primes, it is not hard to see that $ab = (a, b)[a, b]$: if

$$a = p_1^{e_1} p_2^{e_2} \cdots p_k^{e_k} \quad \text{and} \quad b = p_1^{f_1} p_2^{f_2} \cdots p_k^{f_k}$$

then

$$(a, b) = p_1^{g_1} p_2^{g_2} \cdots p_k^{g_k} \quad \text{and} \quad [a, b] = p_1^{h_1} p_2^{h_2} \cdots p_k^{h_k}$$

where $g_i = \min(e_i, f_i)$ and $h_i = \max(e_i, f_i)$. Because

$$e_i + f_i = \min(e_i, f_i) + \max(e_i, f_i) = g_i + h_i$$

the result follows. However, it is possible to get it without using that theorem:

Theorem *If a and b are positive integers, then $(a, b)[a, b] = ab$.*

Proof Let $d = (a, b)$, so $a = rd$ and $b = sd$ with $(r, s) = 1$. Put $t = ab/d$. Then

$$t = \frac{rdb}{d} = rb \quad \text{and} \quad t = \frac{asd}{d} = sa.$$

Thus t is a common multiple of a and b. Let u be a common multiple of a and b. Then

$$u = s'a = s'rd \quad \text{and} \quad u = r'b = r'sd$$

so $s'r = r's$. Thus $r \mid r's$ and because $(r, s) = 1$ we have $r \mid r'$. That is,

$r' = wr$ and so

$$u = r'b = wrb = wt.$$

That is, u is a multiple of t. Thus t is the least common multiple of a and b, so

$$[a, b] = \frac{ab}{d},$$

which is what we want.

CHAPTER 2

UNIQUE FACTORIZATION

Though we are not born with it, the knowledge that every positive integer is the product of primes in only one way (except for permutations of the factors) may seem to be so natural as to not be worth commenting on. Nevertheless, we will prove it, if only for completeness' sake.

Definition An integer greater than 1 that has no positive divisors other that itself and 1 is called *prime*. An integer greater than 1 that is not prime is *composite*. The integer 1 is neither prime nor composite: it is a *unit*.

Theorem (Euclid) *There are infinitely many primes.*

Proof Let the first n primes be $p_1, p_2, \ldots, p_n$. Let

$$N = p_1 p_2 \cdots p_n + 1.$$

N is either prime or composite. If it is composite it has a prime divisor. That divisor cannot be any of $p_1, p_2, \ldots, p_n$ because when N is divided by one of them, there is a remainder of 1. So, whether N is prime or composite, there is a prime other than $p_1, p_2, \ldots, p_n$ and hence infinitely many.

Lemma *Every integer greater than* 1 *is divisible by a prime.*

Proof If $n > 1$ is not prime then it has a divisor, d, smaller than n. If d is not prime, then it has a divisor smaller than d. And so on: because a sequence of positive integers cannot decrease forever, eventually we will encounter a prime that divides n.

Lemma *Every integer greater than* 1 *can be written as a product of primes.*

Proof By the first lemma, $n > 1$ has a prime divisor p_1, so $n = n_1 p_1$ with $n_1 < n$. If $n_1 > 1$, it has a prime divisor p_2, so $n = p_1 p_2 n_2$ with $n_2 < n_1$. If $n_2 > 1$ we continue, generating a sequence $n > n_1 > n_2 > \ldots$. Because a sequence of positive integers cannot decrease forever, we will eventually have $n_k = 1$ and $n = p_1 p_2 \cdots p_k$, a product of primes.

So every integer is a product of primes. We now show that the representation is unique, up to the order of the factors.

Lemma *If p is a prime and $p \mid ab$, then $p \mid a$ or $p \mid b$.*

Proof If p is prime then either $(p, a) = p$ or $(p, a) = 1$. In the first case, $p \mid a$. In the second, because we have proved that if $(d, a) = 1$ and $d \mid ab$ then $d \mid b$, we have that $p \mid b$.

In what follows, to the end of the book, p will always denote a prime.

Lemma *If $p \mid a_1 a_2 \ldots a_k$ then $p \mid a_i$ for some i, $i = 1, 2, \ldots, k$.*

Proof Here is an opportunity to use mathematical induction. The lemma is true for $k = 1$. Suppose that it is true for $k = r$. If

$$p \mid a_1 a_2 \cdots a_{r+1} \quad \text{then} \quad p \mid (a_1 a_2 \cdots a_r) a_{r+1}$$

so, by the last lemma,

$$p \mid a_1 a_2 \cdots a_r \quad \text{or} \quad p \mid a_{r+1}.$$

In the first case, if $p \mid a_1 a_2 \cdots a_r$ then, from the induction assumption, $p \mid a_i$ for some i, $i = 1, 2, \ldots, r$. In the second case, $p \mid a_i$ for $i = r + 1$. So in either case, $p \mid a_i$ for some i, $i = 1, 2, \ldots, r + 1$.

Lemma *If $q_1, q_2, \ldots, q_k$ are primes and $p \mid q_1 q_2 \cdots q_k$ then $p = q_i$ for some i.*

Proof This is a corollary of the last lemma. We know that $p \mid q_i$ for some i and because p and q_i are primes, $p = q_i$.

Theorem (Unique Factorization) *Every positive integer can be written as a product of primes in only one way (up to the order of the factors).*

Proof We know that any $n > 1$ can be written as a product of primes. Suppose that this can be done in two ways:

$$n = p_1 p_2 \cdots p_k = q_1 q_2 \cdots q_r.$$

Because $p_1 \mid q_1 q_2 \cdots q_r$ we know that $p_1 \mid q_i$ for some i, $i = 1, 2, \ldots, r$ and hence that $p_1 = q_i$. So that factor may be cancelled, giving

$$p_2 \cdots p_k = q_1 q_2 \cdots q_{i-1} q_{i+1} \cdots q_r.$$

We now have that p_2 is one of the remaining qs, so it may be cancelled. Continuing, we get that each of the ps is one of the qs. When we finish cancelling ps, there can be no qs left over, because if there were, we would have 1 written as a product of primes. So the products contain the same factors, perhaps in different orders. Thus the factorization is unique.

Corollary *Every $n > 1$ can be written in one way (up to order) in the form*

$$n = p_1^{e_1} p_2^{e_2} \cdots p_k^{e_k}$$

with $p_1, p_2, \ldots, p_k$ different primes.

This is the *prime-power decomposition* of n.

Unique factorization does not hold in every set of integers. For example, suppose that we have only the set of integers of the form $4n + 1$, $\{1, 5, 9, 13, \ldots\}$, with multiplication as usual. The product of two integers in the set will be in the set because

$$(4n + 1)(4m + 1) = 4(mn + n + m) + 1.$$

Some members of the set can be written as products of other members, as $25 = 5 \cdot 5$, and others, as 21, cannot. So we have analogies of primes and composites. However, some composites can be written in more than one way as a product of primes, as $693 = 21 \cdot 33 = 9 \cdot 77$. Thus we cannot take unique factorization for granted.

The *Sieve of Eratosthenes* is an ancient method of determining prime numbers. It is based on the

Theorem *If n is composite it has a prime divisor p such that $p \le \sqrt{n}$.*

Proof If n is composite then $n = ab$ for integers a and b, both in the interval $[2, n - 1]$. One of a and b must be less than or equal to $\sqrt{n}$ because if $a > \sqrt{n}$ and $b > \sqrt{n}$ we would have $n = ab > \sqrt{n}\sqrt{n} = n$, which is impossible. If $a \le \sqrt{n}$ then it has a prime divisor (perhaps a itself) that is less than or equal to $\sqrt{n}$, and the same holds for b.

Corollary (the Sieve of Eratosthenes) *List the integers from* 1 *to* N. *For each prime* p, $p \leq \sqrt{N}$, *cross out the integers* $2p, 3p, \ldots$ *that are in the list. The integers that remain are the primes in the interval* $[2, N]$.

Proof If an integer n in the interval $[2, N]$ is composite then it has for a divisor one of the primes p, $p \leq \sqrt{N}$. Thus $n = kp$ for some k and it has been crossed out.

If we have the prime-power decompositions of two integers we can determine their greatest common divisor without using the Euclidean algorithm by picking out the highest power of the primes that divide both. For example, $(1500, 450) = 150$ because

$$(1500, 450) = (2^2 \cdot 3 \cdot 5^3, 2 \cdot 3^2 \cdot 5^2) = 2 \cdot 3 \cdot 5^2 = 150.$$

As noted before,

$$\left(p_1^{e_1} p_2^{e_2} \cdots p_k^{e_k}, p_1^{f_1} p_2^{f_2} \cdots p_k^{f_k}\right) = p_1^{\min(e_1, f_1)} p_2^{\min(e_2, f_2)} \cdots p_k^{\min(e_k, f_k)}$$

and, for least common multiples,

$$\left[p_1^{e_1} p_2^{e_2} \cdots p_k^{e_k}, p_1^{f_1} p_2^{f_2} \cdots p_k^{f_k}\right] = p_1^{\max(e_1, f_1)} p_2^{\max(e_2, f_2)} \cdots p_k^{\max(e_k, f_k)}.$$

(Some of the exponents may be 0.)

CHAPTER 3

Linear Diophantine Equations

A *diophantine equation* (named after Diophantus of Alexandria (c. third century) who was, as far as we know, the first to consider equations with restricted solutions) will be for us one for which we are to find solutions in integers. Some diophantine equations, such as $x^4 + y^4 = 2$ have solutions ($x = y = 1$) and some, such as $x^4 + y^4 = 3$, do not.

A simple diophantine equation is the *linear diophantine equation*, $ax + by = c$. If $(a, b) = 1$ it has infinitely many solutions.

Theorem If $(a, b) = 1$, then all solutions of $ax + by = c$ are

$$x = r + tb, y = s - ta$$

for r and s such that $ar + bs = c$, where t can be any integer.

Proof Because $(a, b) = 1$ we know that there are integers u and v such that $au + bv = 1$. Thus $a(cu) + b(cv) = c$ and we have a solution: $x = r = cu$, $y = s = cv$. That $x = r + tb$, $y = s - ta$ is a solution for every t follows from

$$a(r + tb) + b(s - ta) = (ar + bs) + (abt - abt) = c + 0 = c.$$

Now let (x, y) be any solution of $ax + by = c$. We want to show that $x = r + tb$ and $y = s - ta$ for some integer t.

We have $c - c = (ax + by) - (ar + bs)$ or

$$a(x - r) + b(y - s) = 0. \qquad (1)$$

This implies that $b \mid a(x-r)$ and because $(a, b) = 1$, that $b \mid (x-r)$. That is, there is an integer t such that

$$tb = x - r \quad \text{or} \quad x = r + tb.$$

Substituting in (1) gives

$$atb + b(y - s) = 0 \quad \text{or} \quad y = s - ta.$$

If $(a, b) = d > 1$, then unless $d \mid c$ the equation $ax + by = c$ has no solution because $ax + by$ would be a multiple of d while c is not. If d does divide c, the equation can be turned into

$$\frac{a}{d}x + \frac{b}{d}y = \frac{c}{d}.$$

Then, because $(a/d, b/d) = 1$ the theorem applies and if we can find one solution then we know them all.

For example, let us solve $6x + 21y = 111$. Divide by the common factor to get $2x + 7y = 37$. Inspection, or trial, gives $x = 1$ and $y = 5$ as a solution. We then know that all solutions are

$$x = 1 + 7t, \quad y = 5 - 2t$$

where t is an integer. We will later get methods other than trial or inspection for finding a particular solution.

CHAPTER 4

CONGRUENCES

The invaluable congruence notation was devised by Gauss.

Definition $a \equiv b \pmod{m}$ ("a is congruent to b modulo m") if and only if $m \mid (a - b)$.

We will always suppose that m is positive.

For example, $7 \equiv 17 \pmod{10}$, and another way of saying "n is even" is "$n \equiv 0 \pmod{2}$".

The definition has some immediate consequences that, for want of a better word, we will label as theorems.

Theorem *$a \equiv b \pmod{m}$ if and only if there is an integer k such that $a = b + km$.*

Proof If $a \equiv b \pmod{m}$ then $m \mid (a - b)$. The definition of divisibility says that there is an integer k such that $km = ab$, so $a = b + km$. Conversely, if $a = b + km$, then $m \mid (a - b)$ and so $a \equiv b \pmod{m}$.

For example, if an integer n has last digit 2, then $n = 2 + 10k$ for some k and $n \equiv 2 \pmod{10}$.

Theorem *Every integer is congruent* (mod m) *to one of* $0, 1, \ldots, m - 1$.

Proof Given a, the division algorithm says that there are integers q and r, $0 \leq r \leq m - 1$, such that $a = qm + r$. Then $a \equiv r \pmod{m}$.

Definition The integer r, $0 \leq r \leq m - 1$, such that $a \equiv r \pmod{m}$ is the *least residue of a* (mod m).

The least residue of 1937 (mod 10) is 7, and its least residue (mod 11) is 1, because $1937 = 1 + 176 \cdot 11$.

Theorem *$a \equiv b \pmod{m}$ if and only if a and b leave the same remainder on division by m.*

Proof If $a/m = u + r/m$ and $b/m = v + r/m$ then $a/m - u = b/m - v$ so $a - b = m(u - v)$. From the definition of congruence, $a \equiv b \pmod{m}$. Conversely, if $a \equiv b \pmod{m}$, then $a \equiv b \equiv r \pmod{m}$ where r is a least residue (mod m) and hence $0 \leq r \leq m - 1$. Thus $a = um + r$ and $b = vm + r$ and so have the same remainder when divided by m.

Each of the three ways of looking at $a \equiv b \pmod{m}$:

$$m \mid (a - b),$$

$$a = b + km, \text{ and}$$

a and b leave the same remainder on division by m

is useful in one place or another.

Congruence shares many properties with equality.

Theorem *For integers a, b, c, and d and modulus m,*

(a) $a \equiv a \pmod{m}$,

(b) *if* $a \equiv b \pmod{m}$ *then* $b \equiv a \pmod{m}$,

(c) *if* $a \equiv b \pmod{m}$ *and* $b \equiv c \pmod{m}$ *then* $a \equiv c \pmod{m}$,

(d) *if* $a \equiv b \pmod{m}$ *and* $c \equiv d \pmod{m}$ *then* $a + c \equiv b + d \pmod{m}$,

(e) *if* $a \equiv b \pmod{m}$ *and* $c \equiv d \pmod{m}$ *then* $ac \equiv bd \pmod{m}$.

Proof All parts follow directly from the definition of congruence. For example, for (d), we know that $a = b + km$ and $c = d + jm$ for some integers j and k. Then

$$a + c = b + d + m(j + k) \quad \text{so} \quad a + c \equiv b + d \pmod{m}.$$

The difference comes in division. In equalities we can cancel non-zero factors but in congruences we can do this only if the factor and the modulus are relatively prime.

Theorem *If $ac \equiv bc \pmod{m}$ and $(c, m) = 1$, then $a \equiv b \pmod{m}$.*

Proof From the definition of congruence, $m \mid (ac - bc)$ or $m \mid c(a - b)$. From the theorem that if $x \mid yz$ and $(x, y) = 1$ then $x \mid z$, we conclude that $m \mid (a - b)$ and $a \equiv b \pmod{m}$.

If the factor and the modulus are not relatively prime, we have the

Theorem *If $ac \equiv bc \pmod{m}$ and $(c, m) = d$, then $a \equiv b \pmod{m/d}$.*

Proof We have $ac - bc = km$ so $(a - b)(c/d) = k(m/d)$. That is, $a(c/d) \equiv b(c/d) \pmod{m/d}$. Because $(c/d, m/d) = 1$ the last theorem says that $a \equiv b \pmod{m/d}$.

So, if $5x \equiv 25 \pmod{100}$, then $x \equiv 5 \pmod{20}$.

As an application of congruences, we show that none of 7, 15, 23, ..., that is, no integer congruent to 7 (mod 8), is a sum of three squares of integers. Modulo 8, every integer is congruent to one of 0, 1, 2, 3, 4, 5, 6, or 7. So every square is congruent (mod 8) to one of 0, 1, 4, 9, 16, 25, 36, or 49. Taking least residues (mod 8), every square (mod 8) is 0, 1, 4, 1, 0, 1, 4, or 1. Every sum of three squares will be (mod 8) the sum of three of these. There is no way to take three integers, each 0, 1, or 4, and add them to get a sum of 7 (mod 8). (0, 1, 2, 3, 4, 5, and 6 are possible, but not 7.) Thus no $n \equiv 7 \pmod 8$ is a sum of three squares.

Congruences can be helpful if we need to solve a linear diophantine equation. For example, suppose we want to solve $6x + 21y = 111$. Divide out the common factor to get $2x + 7y = 37$. Modulo 7, we have $2x \equiv 37 \equiv 2 \pmod 7$, so $x \equiv 1 \pmod 7$. That is, $x = 1 + 7t$. Substitute: $2(1 + 7t) + 7y = 37$. This becomes $14t + 7y = 35$ or $2t + y = 5$. We have found all solutions: $x = 1 + 7t$, $y = 5 - 2t$.

In the chapter on continued fractions, we will see how the continued fraction expansion of a/b can be used to solve $ax + by = 1$.

CHAPTER 5

LINEAR CONGRUENCES

When confronted with the linear equation $ax = b$ we say "$x = b/a$" and pass on. The linear congruence $ax \equiv b \pmod{m}$ is a little more complicated.

For one thing, if we have one integer that satisfies it we have infinitely many, because we can add to or subtract from it any multiple of m. We single one number out:

Definition A *solution* to $ax \equiv b \pmod{m}$ is an integer that satisfies the congruence that is a least residue (mod m). (That is, one of $0, 1, \ldots, m-1$.)

For example, $2x \equiv 3 \pmod{7}$ is true for $x = \ldots, -9, -2, 5, 12, 19, \ldots$ but its solution is 5.

We will work up to the proof of the

Theorem *If b is not a multiple of (a, m) then $ax \equiv b \pmod{m}$ has no solutions. If b is a multiple of (a, m), then $ax \equiv b \pmod{m}$ has (a, m) solutions. In particular, if $(a, m) = 1$, then $ax \equiv b \pmod{m}$ has a unique solution.*

We first take care of the case of no solutions.

Lemma *If b is not a multiple of (a, m) then $ax \equiv b \pmod{m}$ has no solutions.*

Proof The congruence can be written as $ax = b + km$ for some k, or $b = ax - km$. The right-hand side is a multiple of (a, m). If the left-hand side is not, the equation is impossible.

Thus we see by inspection that $4x \equiv 297 \pmod{23592}$ has no solutions. Next is the case of one solution.

Lemma *If* $(a, m) = 1$, *then* $ax \equiv b \pmod{m}$ *has exactly one solution.*

Proof We have previously proved the theorem that the greatest common divisor of two integers can be written as a linear combination of them. So we know that there are integers r and s such that $ra + sm = 1$. Thus

$$a(rb) + (sb)m = b$$

so rb satisfies $ax \equiv b \pmod{m}$. Its least residue (mod m) will be a solution.

Suppose that there are two solutions u and v, so $au \equiv av \equiv b \pmod{m}$. Then $a(u - v) \equiv 0 \pmod{m}$. Because $(a, m) = 1$ this implies $u - v \equiv 0 \pmod{m}$. But u and v are least residues (mod m), so they are each one of $0, 1, \ldots, m - 1$ and hence

$$-(m - 1) \leq u - v \leq m - 1.$$

The only number congruent to 0 (mod m) in that range is 0, so $u = v$. So there are not two solutions, but only one.

For example, $4x \equiv 297 \pmod{23593}$ has exactly one solution. Finding it by trial, which will always work, might take some time, but if it is really needed continued fractions (see Chapter 27) can disclose it with less work.

We almost have the general case.

Lemma *Let* $d = (a, m)$. *If* $d \mid b$, *then* $ax \equiv b \pmod{m}$ *has exactly* d *solutions.*

Proof We know that we can write the congruence as

$$(a/d)x \equiv b/d \pmod{m/d}.$$

Because $(a/d, m/d) = 1$ we know from the last lemma that the congruence has exactly one solution, call it r. Let s be any other solution of $ax \equiv b \pmod{m}$. Then $as \equiv ar \pmod{m}$ and we know (see a theorem on page 16) that this gives $s \equiv r \pmod{m/d}$. Thus $s = r + k(m/d)$ for some k. Putting $k = 0, 1, \ldots, d-1$ we get solutions that are least residues (mod m) because, r being a least residue (mod m/d) and hence less than m/d,

$$0 \leq r + k\frac{m}{d} \leq r + (d-1)\frac{m}{d} < \frac{m}{d} + (d-1)\frac{m}{d} = m.$$

All such integers satisfy $ax \equiv b \pmod{m}$ because

$$\frac{a}{d}\left(r + k\frac{m}{d}\right) \equiv \frac{a}{d}r \equiv \frac{b}{d} \pmod{m/d}$$

and this implies

$$a\left(r + k\frac{m}{d}\right) \equiv b \pmod{m}.$$

Proof of the Theorem The three lemmas combine to provide a proof.

Solutions to $ax \equiv b \pmod{m}$ with $(a, m) = 1$ may be found by trial (one of $1, 2, \ldots, m - 1$ must satisfy the congruence) or by the exercise of more or less ingenuity. For example, to solve

$$17x \equiv 28 \pmod{39},$$

we may turn it into

$$-22x \equiv 28 \pmod{39} \quad \text{or} \quad -11x \equiv 14 \pmod{39}.$$

Adding that to the original congruence gives

$$6x \equiv 42 \pmod{39} \quad \text{or} \quad x \equiv 7 \pmod{13}.$$

The original congruence has exactly one solution: $x = 7$ does not work but $x = 20$ does. So we do not need to check $x = 33$.

CHAPTER 6

THE CHINESE REMAINDER THEOREM

Some ancient Chinese manuscripts posed the problem of determining an integer given its remainders on division by other integers, whence the name of the theorem. Its statement is

Theorem *The system of congruences* $x \equiv a_i \pmod{m_i}$, $i = 1, 2, \ldots, k$, *where the moduli are pairwise relatively prime (i.e.,* $(m_i, m_j) = 1$ *if* $i \neq j$*) has a unique solution modulo* $m_1 m_2 \cdots m_k$.

There are two standard proofs, the first of which generalizes what is done when the system is solved in the obvious manner, as in the

Example Let us solve

$$x \equiv 1 \pmod 3, \quad x \equiv 2 \pmod 5, \quad x \equiv 3 \pmod 7.$$

The first congruence says that $x = 1 + 3r$ for some r. Substitute this into the second congruence: r must satisfy $1 + 3r \equiv 2 \pmod 5$. This gives $r \equiv 2 \pmod 5$, or $r = 2 + 5s$ for some s. Thus

$$x = 1 + 3r = 1 + 3(2 + 5s) = 7 + 15s$$

satisfies the first two congruences. To satisfy the third, we must have $7 + 15s \equiv 3 \pmod 7$. This gives $s \equiv 3 \pmod 7$, or $s = 3 + 7t$. Thus

$$x = 7 + 15s = 7 + 15(3 + 7t) = 52 + 105t$$

satisfies all three congruences, so $x \equiv 52 \pmod{105}$ is the solution.

This can be continued for any number of congruences, an idea that is made formal in the

Proof We use mathematical induction. The theorem is true when $k = 1$. Suppose that it true for $k = r$. Then we know that $x \equiv a_i \pmod{m_i}$, $i = 1, 2, \ldots, r$ has a unique solution, call it S, modulo $m_1 m_2 \cdots m_r$. To have a solution to $x \equiv a_i \pmod{m_i}$, $i = 1, 2, \ldots, r+1$, x must satisfy

$$x \equiv S \pmod{m_1 m_2 \cdots m_r}$$
$$x \equiv a_{r+1} \pmod{m_{r+1}}.$$

We know that $x = S + u m_1 m_2 \cdots m_r$ for some u, so we need to find u such that

$$S + u m_1 m_2 \cdots m_r \equiv a_{r+1} \pmod{m_{r+1}}.$$

Because $(m_1 m_2 \cdots m_r, m_{r+1}) = 1$ (if $p \mid m_1 m_2 \cdots m_r$ and $p \mid m_{r+1}$ then p is a common factor of two of the moduli) this linear congruence has exactly one solution. Substituting the value of u in $x = S + u m_1 m_2 \cdots m_r$ gives a solution to both congruences modulo $m_1 m_2 \cdots m_r m_{r+1}$.

It remains to show that the solution is unique. Suppose that there are two solutions, so $r \equiv s \equiv a_i \pmod{m_i}$, $i = 1, 2, \ldots, k$. Then $m_i \mid (r - s)$ and $r - s$ is a common multiple of all the m_i. Thus it is a multiple of their least common multiple. Because the moduli are relatively prime in pairs, the least common multiple is $m_1 m_2 \cdots m_k$ and so $r - s \equiv 0 \pmod{m_1 m_2 \cdots m_k}$. Since r and s are least residues (mod $m_1 m_2 \cdots m_k$), we have $r = s$.

There is another standard proof, shorter because more clever:

Alternate proof Let $M = m_1 m_2 \cdots m_k$. Because $(M/m_i, m_i) = 1$ we know that there is a solution to each of the linear congruences

$$(M/m_i)x \equiv a_i \pmod{m_i}, \quad i = 1, 2, \ldots, k.$$

Call the solution to the ith congruence s_i and let

$$S = \frac{M}{m_1} s_1 + \frac{M}{m_2} s_2 + \cdots + \frac{M}{m_k} s_k.$$

Because

$$m_i \left| \frac{M}{m_j} \right.$$

if $i \neq j$ when we look at the equation (mod m_i) all the terms except one drop out, leaving

$$S \equiv \frac{M}{m_i} s_i \equiv a_i \pmod{m_i},$$

so S satisfies all of the congruences. To show that the solution is unique, we do the same as we did in the first proof.

In the example, we would solve the congruences $35x \equiv 1 \pmod{3}$, $21x \equiv 2 \pmod{5}$, and $15x \equiv 3 \pmod{7}$, getting the solutions 2, 2, and 3 and then combine them to get

$$S = 35 \cdot 2 + 21 \cdot 2 + 15 \cdot 3 = 157,$$

leading to the same solution, 52 (mod 105), as before.

CHAPTER 7

Fermat's Theorem

Fermat's theorem, sometimes called "Fermat's Little Theorem" to distinguish it from Fermat's Last Theorem about solutions of $x^n + y^n = z^n$, is useful in many places.

Theorem *If p is prime and $(a, p) = 1$, then $a^{p-1} \equiv 1 \pmod{p}$.*

Though hardly deserving the name of lemma—"observation" would be more appropriate—we need the

Lemma *If $(a, m) = 1$, then the least residues of $a, 2a, \ldots, (m-1)a$ $\pmod{m}$ are a permutation of $1, 2, \ldots, m-1$.*

For example, for $m = 8$ and $a = 3$ the two sets are 3, 6, 9, 12, 15, 18, 21 and 3, 6, 1, 4, 7, 2, 5.

Proof There are $m - 1$ multiples of a, each congruent to one of $1, 2, \ldots, m-1 \pmod{a}$, and $m - 1$ integers from 1 to $m - 1$, so to prove the lemma it suffices to show that the least residues of the multiples of a are all different.

Suppose that $ra \equiv sa \pmod{m}$. Because $(a, m) = 1$ we know by a previous theorem that $r \equiv s \pmod{m}$. (The previous theorem is that if $m \mid ax$ and $(m, a) = 1$, then $m \mid x$, applied to $m \mid (ra - sa)$.) That is, $m \mid (r - s)$. But r and s both lie in the interval $[1, m-1]$, so $-(m-1) < r - s < m - 1$. The only multiple of m in that interval is 0, so $r - s = 0$ and $r = s$. Thus two different multiples cannot be congruent $\pmod{m}$.

Proof of Fermat's Theorem The lemma tells us that for any prime p if $(a, p) = 1$, then the least residues $\pmod{p}$ of $a, 2a, \ldots, (p-1)a$ are a permutation of $1, 2, \ldots, p-1$. So, when we multiply them together, we have

$$a \cdot 2a \cdots (p-1)a \equiv 1 \cdot 2 \cdots (p-1) \pmod{p}$$

or

$$a^{p-1}(p-1)! \equiv (p-1)! \pmod{p}.$$

Because $(p, (p-1)!) = 1$, this gives $a^{p-1} \equiv 1 \pmod{p}$, which is Fermat's theorem.

The theorem may be stated slightly differently:

Corollary *If p is a prime, then $a^p \equiv a \pmod{p}$ for all a.*

Proof If $(a, p) = 1$ this follows from Fermat's theorem by multiplication. If $(a, p) = p$, it says that $0 \equiv 0 \pmod{p}$, which is true. There are no other cases.

As a not terribly important application of Fermat's theorem we have the

Theorem *Every prime other than 2 or 5 divides some integer whose digits are all 1s.*

Proof If $p \neq 2, 3$, or 5, then, because of Fermat's Theorem, p divides

$$\frac{10^{p-1} - 1}{9},$$

which is $111 \ldots 11$ ($(p-2)$ digits, all 1s). Noting that $3 \mid 111$ completes the proof.

CHAPTER 8

WILSON'S THEOREM

Wilson's theorem was not proved by Wilson, but by Lagrange in 1770.

Theorem *An integer p, $p \geq 2$, is prime if and only if* $(p-1)! \equiv -1 \pmod{p}$.

For example, 5 is prime and $4! = 24 \equiv -1 \pmod{5}$ whereas 6 is composite and $5! = 120 \equiv 0 \pmod{6}$.

The theorem gives a necessary and sufficient condition for an integer to be a prime, but it does not provide a practical primality test. It is, however, useful. For example, a polynomial has been constructed whose positive integer values are, for integer values of the variables, all the primes. What makes it work is, fundamentally, Wilson's theorem. (Unfortunately for computation, its values are almost always negative.)

The idea of the proof is similar to that of Fermat's theorem. We will break it up with two preliminary lemmas.

Lemma *If $p > 2$ is prime, then the congruence $x^2 \equiv 1 \pmod{p}$ has exactly two solutions, 1 and $p-1$.*

Proof Let r be any solution. (As always, by "solution" to a congruence $\pmod{p}$ we mean a least residue, an integer from 0 to $p-1$, that satisfies the congruence.) Then $r^2 - 1 \equiv 0 \pmod{p}$ so $p \mid (r+1)(r-1)$. Because p is prime, either $p \mid (r+1)$ or $p \mid (r-1)$. That is, $r \equiv -1 \pmod{p}$ or $r \equiv 1 \pmod{p}$. Because r is a least residue, $r = p-1$ or $r = 1$.

If p is not prime, the lemma is false. For example, $x^2 \equiv 1 \pmod{8}$ has four solutions, $x = 1, 3, 5, 7$.

If $(a, p) = 1$, we know that $ax \equiv 1 \pmod{p}$ has exactly one solution. It can be thought of as the reciprocal of $a \pmod{p}$ and, as could be expected and as we now show, it is unique. For example, for $p = 13$ we have the reciprocals (mod 13)

a	1	2	3	4	5	6	7	8	9	10	11	12
a'	1	7	9	10	8	11	2	5	3	4	6	12.

Lemma *Let p be an odd prime and let a' denote the solution of $ax \equiv 1 \pmod{p}$. Then $a' \equiv b' \pmod{p}$ if and only if $a \equiv b \pmod{p}$. Furthermore, $a \equiv a' \pmod{p}$ if and only if $a \equiv \pm 1 \pmod{p}$.*

Proof Suppose that $a' \equiv b' \pmod{p}$. Then

$$b \equiv aa'b \equiv ab'b \equiv a \pmod{p}.$$

Conversely, suppose that $a \equiv b \pmod{p}$. Then

$$b' \equiv b'aa' \equiv b'ba' \equiv a' \pmod{p}.$$

For the second part of the lemma, from $1 \cdot 1 \equiv (p-1)(p-1) \equiv 1 \pmod{p}$ it follows that $1' = 1$ and $(p-1)' = p-1$. Conversely, if $a \equiv a' \pmod{p}$ then $1 \equiv aa' \equiv a^2 \pmod{p}$ and from the first lemma, $a = 1$ or $p-1$, which is what we want.

Proof of Wilson's theorem Suppose that p is prime. If $p = 2$ or 3, the theorem is true. If $p > 3$, then from the last lemma we can group the integers $2, 3, \ldots, p-2$ into pairs consisting of an integer a and its reciprocal a', which is different from a. For example, if $p = 13$, we have the pairs $(2, 7)$, $(3, 9)$, $(4, 10)$, $(5, 8)$, and $(6, 11)$. Because the product of the integers in each pair is 1 (mod p), we have

$$2 \cdot 3 \cdots (p-2) \equiv 1 \pmod{p}.$$

Hence

$$(p-1)! \equiv 1 \cdot 2 \cdot 3 \cdots (p-2)(p-1) \equiv 1 \cdot 1 \cdot (p-1) \equiv -1 \pmod{p}.$$

We have proved half the theorem. It remains to show that if $(n-1)! \equiv -1 \pmod{n}$ then n is prime. Suppose not. Then $n = ab$ for some integers a and b in the interval $[2, n-1]$. We have supposed that $n \mid ((n-1)! + 1)$. Because $a \mid n$ we know that $a \mid ((n-1)! + 1)$. But because $a \leq n-1$, one of the factors of $(n-1)!$ is a itself. Thus $a \mid (n-1)!$. Because a divides both $(n-1)!$ and $(n-1)! + 1$, it follows that $a \mid 1$, which is a contradiction. Thus n is prime.

If n is composite then $(n-1)! \equiv 0 \pmod{n}$ for $n \geq 6$ because $(n-1)!$ contains all the factors of n.

CHAPTER 9

THE NUMBER OF DIVISORS OF AN INTEGER

To keep statements uncomplicated, in this chapter and the next three all integers will be positive.

Let $d(n)$ denote the number of positive divisors of n. For example, because 60 has divisors 1, 2, 3, 4, 5, 6, 10, 12, 15, 20, 30, and 60, we have $d(60) = 12$.

We can calculate $d(n)$ from the prime-power decomposition of n.

Theorem *If*

$$n = p_1^{e_1} p_2^{e_2} \cdots p_k^{e_k}$$

then

$$d(n) = (e_1 + 1)(e_2 + 1) \cdots (e_k + 1).$$

Proof Any divisor of n has the form

$$n = p_1^{a_1} p_2^{a_2} \cdots p_k^{a_k}$$

where $0 \le a_i \le e_i$ for $i = 0, 1, \ldots, e_i$. Each a_i may be chosen in $e_i + 1$ ways, so the total number of divisors is $(e_1 + 1)(e_2 + 1) \cdots (e_k + 1)$.

Thus, without counting divisors we see that

$$d(60) = d(2^2 \cdot 3 \cdot 5) = 3 \cdot 2 \cdot 2 = 12.$$

A class of functions to which d belongs is that of multiplicative functions.

Definition A function f is *multiplicative* if $(m, n) = 1$ implies $f(mn) = f(m)f(n)$.

Theorem *d is multiplicative.*

Proof If m and n are relatively prime, then the prime-power decompositions of m and n have no primes in common. That is, their prime-power decompositions are

$$m = p_1^{e_1} p_2^{e_2} \cdots p_k^{e_k} \quad \text{and} \quad n = q_1^{f_1} q_2^{f_2} \cdots q_r^{f_r}$$

where none of the ps equals any of the qs and vice versa. Thus

$$\begin{aligned} d(mn) &= (e_1+1)(e_2+1)\cdots(e_k+1)(f_1+1)(f_2+1)\cdots(f_r+1) \\ &= d(m)d(n). \end{aligned}$$

A property of multiplicative functions is that their values on prime powers determine their values on all integers.

Theorem *If f is multiplicative and the prime-power decomposition of n is $n = p_1^{e_1} p_2^{e_2} \cdots p_k^{e_k}$, then*

$$f(n) = f(p_1^{e_1}) f(p_2^{e_2}) \cdots f(p_k^{e_k}).$$

Proof The proof is by induction on k. The theorem is true for $k = 1$. Suppose that it is true for $k = r$. Because $(p_1^{e_1} p_2^{e_2} \cdots p_r^{e_r}, p_{r+1}^{e_{r+1}}) = 1$ the definition of multiplicative function gives us

$$f\left((p_1^{e_1} p_2^{e_2} \cdots p_r^{e_r}) p_{r+1}^{e_{r+1}}\right) = f(p_1^{e_1} p_2^{e_2} \cdots p_r^{e_r}) f(p_{r+1}^{e_{r+1}}).$$

From the induction assumption, the first factor is

$$f(p_1^{e_1}) f(p_2^{e_2}) \cdots f(p_r^{e_r}).$$

That, together with the preceding equation, completes the induction.

For example, if we know that f is multiplicative and $f(p^n) = np^{n-1}$ then we can calculate the value of f for any n, as in

$$f(600) = f(2^3 \cdot 3 \cdot 5^2) = f(2^3) f(3) f(5^2) = (3 \cdot 2^2)(1)(2 \cdot 5) = 120.$$

CHAPTER 10

THE SUM OF THE DIVISORS OF AN INTEGER

Let $\sigma(n)$ denote the sum of the divisors of n. For example,

$$\sigma(30) = 1 + 2 + 3 + 5 + 6 + 10 + 15 + 30 = 72.$$

As we will see shortly, this can be obtained more quickly from

$$\sigma(30) = \sigma(2 \cdot 3 \cdot 5) = \sigma(2)\sigma(3)\sigma(5) = 3 \cdot 4 \cdot 6 = 72.$$

For a prime p,

$$\sigma(p^n) = 1 + p + p^2 + \cdots + p^n.$$

This is enough for us to determine $\sigma(n)$ for all n as is shown by the following theorem.

Theorem *If $p_1^{e_1} p_2^{e_2} \cdots p_k^{e_k}$ is the prime-power decomposition of n, then*

$$\sigma(n) = \sigma(p_1^{e_1})\sigma(p_2^{e_2}) \cdots \sigma(p_k^{e_k}).$$

Proof We will use mathematical induction. The theorem is true for $k = 1$. Suppose that it is true for $k = r$. We want to show that this implies its truth for $k = r + 1$. Let

$$n = p_1^{e_1} p_2^{e_2} \cdots p_r^{e_r} p_{r+1}^{e_{r+1}} = N p_{r+1}^{e_{r+1}}.$$

To simplify the notation, we will write $n = Np^e$. Let the divisors of N be $1, d_1, \ldots, d_t$. Because $(N, p) = 1$, all the divisors of n are

$$\begin{array}{ccccc} 1 & d_1 & d_2 & \cdots & d_t \\ p & d_1 p & d_2 p & \cdots & d_t p \\ & & & \vdots & \\ p^e & d_1 p^e & d_2 p^e & \cdots & d_t p^e . \end{array}$$

Summing, we get

$$\begin{aligned} \sigma(n) &= (1 + d_1 + \cdots + d_t)(1 + p + \cdots + p^e) \\ &= \sigma(N)\sigma(p^e) \end{aligned}$$

From the induction assumption, $\sigma(N) = \sigma(p_1^{e_1})\sigma(p_2^{e_2})\cdots\sigma(p_k^{e_k})$ and the last two equations complete the proof.

Another way to write the theorem, which may seem more compact but which says the same thing, is the

Corollary *If the prime-power decomposition of n is $n = p_1^{e_1} p_2^{e_2} \cdots p_k^{e_k}$ then*

$$\sigma(n) = \frac{p_1^{e_1+1} - 1}{p_1 - 1} \cdot \frac{p_2^{e_2+1} - 1}{p_2 - 1} \cdots \frac{p_1^{e_k+1} - 1}{p_k - 1}.$$

Proof For each i,

$$\sigma(p_i^{e_i}) = 1 + p_i + p_i^2 + \cdots + p_i^{e_i} = \frac{p_i^{e_i+1} - 1}{p_i - 1}.$$

As with the number-of-divisors function d, we have the

Theorem *σ is multiplicative.*

Proof The proof is the same as that for d in the last chapter, replacing d with σ.

A generalization of $\sigma(n)$ is $\sigma_k(n)$, the sum of the kth powers of the divisors of n. When $k = 0$, the function is $d(n)$ and when $k = 1$ it is $\sigma(n)$. As with d and σ, it is multiplicative and because

$$\sigma_k(p^n) = 1 + p^k + p^{2k} + \cdots + p^{nk} = \frac{1 - p^{(n+1)k}}{1 - p^k}$$

its value can be found for any integer.

CHAPTER 11

AMICABLE NUMBERS

Two integers are *amicable* if the sum of the divisors of one, excluding the number itself, is the other, and vice versa. That is, m and n are amicable if $\sigma(m) - m = n$ and $\sigma(n) - n = m$ or if

$$\sigma(m) = \sigma(n) = m + n.$$

For example, the smallest pair of amicable numbers, $(220, 284)$, is such because

$$\sigma(220) = \sigma(2^2 \cdot 5 \cdot 11) = \sigma(2^2)\sigma(5)\sigma(11) = 7 \cdot 6 \cdot 12 = 504$$

and

$$\sigma(284) = \sigma(2^2 \cdot 71) = \sigma(2^2)\sigma(71) = 7 \cdot 72 = 504 = 220 + 284.$$

The name dates back to ancient Greek number mysticism. (The ancient Greeks, lacking the σ function, would have verified the amicability of 220 and 284 by calculating

$$1 + 2 + 4 + 5 + 10 + 11 + 20 + 22 + 44 + 55 + 110 = 284$$

and

$$1 + 2 + 4 + 71 + 142 = 220.)$$

Amicable numbers belong more to recreational mathematics than to number theory, though Euler once thought it worth his while to discover sixty new amicable pairs. There is a conjecture, widely believed, that there are infinitely many amicable pairs. A few years ago there was hope that it would be proved by finding methods of generating larger amicable pairs from known ones, but the hope has not yet been realized.

Here is one result on amicable pairs.

Theorem *Let $p = 3 \cdot 2^e - 1$, $q = 3 \cdot 2^{e-1} - 1$, and $r = 3^2 \cdot 2^{2e-1} - 1$. If p, q, and r are odd primes, then $m = 2^e pq$ and $n = 2^e r$ are amicable.*

Proof

$$\begin{aligned}\sigma(m) &= \sigma(2^e pq) \\ &= \sigma(2^e)\sigma(p)\sigma(q) \\ &= (2^{e+1} - 1)(p + 1)(q + 1) \\ &= (2^{e+1} - 1)(3 \cdot 2^e)(3 \cdot 2^{e-1}) \\ &= 3^2 \cdot 2^{2e-1}(2^{e+1} - 1)\end{aligned}$$

and

$$\begin{aligned}\sigma(n) &= \sigma(2^e r) \\ &= \sigma(2^e)\sigma(r) \\ &= (2^{e+1} - 1)(r + 1) \\ &= (2^{e+1} - 1) \cdot 3^2 \cdot 2^{2e-1}\end{aligned}$$

so $\sigma(m) = \sigma(n)$. Also,

$$\begin{aligned}m + n &= 2^e pq + 2^e r \\ &= 2^e\big((3 \cdot 2^e - 1)(3 \cdot 2^{e-1} - 1) + 3 \cdot (2^{2e-1} - 1)\big)\end{aligned}$$

That this has the same value as $\sigma(m)$ and $\sigma(n)$ is an exercise in algebra.

When $e = 2$, we get $p = 11$, $q = 5$, and $r = 71$, giving the amicable pair 220, 284. Unfortunately, there are not many values of e for which p, q, and r are all odd primes. The only others may be and $e = 4$ and $e = 7$.

CHAPTER 12

PERFECT NUMBERS

An integer is called a *perfect number* if is equal to the sum of its positive divisors, not including itself. Thus $6 = 1 + 2 + 3$ and $28 = 1 + 2 + 4 + 7 + 14$ are perfect numbers. The name dates back to ancient Greek number mysticism, in which, for example, even numbers were female (because they were easily divided into two parts, so they were weak and hence female—that is the way that ancient Greek number mystics thought). Regardless of their name, perfect numbers are of mathematical interest.

The first recorded theorem on perfect numbers appears in Euclid's *Elements*:

Theorem *If* $2^k - 1$ *is prime, then* $2^{k-1}(2^k - 1)$ *is perfect.*

Proof The sum of the positive divisors of an integer, excluding itself, is $\sigma(n) - n$ so to show that an integer is perfect we need to verify that $\sigma(n) - n = n$, or $\sigma(n) = 2n$. Let $n = 2^{k-1}(2^k - 1)$. Because $2^k - 1$ is prime, $\sigma(2^k - 1) = (2^k - 1) + 1 = 2^k$. We know that σ is multiplicative, so that if $(r, s) = 1$ then $\sigma(rs) = \sigma(r)\sigma(s)$. Applying that to $2^k - 1$ and 2^{k-1}, which have a greatest common divisor of 1, and using

$$\sigma(2^{k-1}) = 1 + 2 + 2^2 + \cdots + 2^{k-1} = 2^k - 1$$

we have

$$\begin{aligned} \sigma(n) &= \sigma\big(2^{k-1}(2^k - 1)\big) \\ &= \sigma(2^{k-1})\sigma(2^k - 1) \\ &= (2^k - 1) \cdot 2^k \\ &= 2n, \end{aligned}$$

so n is perfect.

The first few perfect numbers, corresponding to $k = 2, 3, 5, 7$, and 13, are 6, 28, 496, 8128, and 33550336.

Euler showed that the numbers of the theorem are the only even perfect numbers:

Theorem *If n is an even perfect number, then $n = 2^{p-1}(2^p - 1)$ where p and $2^p - 1$ are primes.*

Proof If n is an even perfect number, then $n = 2^e m$, where m is odd and $e \geq 1$. We know that $\sigma(m) > m$, so $\sigma(m) = m + s$ with $s > 0$. Because n is perfect, $2n = \sigma(n)$, which becomes

$$\begin{aligned} 2(2^e m) &= \sigma(2^e m) \\ &= \sigma(2^e)\sigma(m) \\ &= (2^{e+1} - 1)(m + s) \end{aligned}$$

or

$$2^{e+1}m = 2^{e+1}m + 2^{e+1}s - ms$$

so

$$m = (2^{e+1} - 1)s.$$

This says that s is a divisor of m and $s < m$. But $\sigma(m) = m + s$, so that s is the sum of *all* the divisors of m that are less than m. The only way that can happen is for the sum to contain one element alone, so s is the only divisor of m that is less than m. That implies that $s = 1$, for, if not, m would have more than one divisor less than itself.

Thus $\sigma(m) = m + 1$, which says that m is prime. From the last equation, $m = 2^{e+1} - 1$. The only numbers of this form that can be prime are those with $e + 1$ prime. ($2^{ab} - 1$ is divisible by $2^a - 1$ and by $2^b - 1$.) Thus $e + 1 = p$ is a prime and the prime $m = 2^p - 1$.

No odd perfect numbers are known. Many necessary conditions for an integer to be odd and perfect have been derived that show that, if one exists, it must be very large. No proof that there are no odd perfect numbers is on the horizon. It may be that there are odd perfect numbers but they are too large for us ever to find them.

CHAPTER 13

EULER'S THEOREM AND FUNCTION

After Fermat's Theorem that $a^{p-1} \equiv 1 \pmod{p}$ if p is a prime and $(a, p) = 1$, it is natural to ask if there is a similar theorem when the modulus is not prime. There is.

Definition (Euler's ϕ-function) Let $\phi(n)$ denote the number of positive integers that are less than or equal to n and relatively prime to n.

For example, $\phi(12) = 4$ because the integers from 1 to 12 that are relatively prime to 12 are 1, 5, 7, and 11.

If $(a, m) \neq 1$ then $a^r \equiv 1 \pmod{m}$ is impossible because the congruence is that same as the equation $a^r = 1 + km$ for some k, or $1 = a^r - km$. The greatest common divisor of a and m divides the right-hand side of the last equation and it hence divides the left-hand side, 1. So it has to be 1.

Theorem *If* $(a, m) = 1$, *then* $a^{\phi(m)} \equiv 1 \pmod{m}$.

Proof The idea of the proof is the same as that in the proof of Fermat's theorem, but instead of considering the product of the positive integers less than p we will look at the product of the positive integers less than m and relatively prime to m.

We show first that if $r_1, r_2, \ldots, r_{\phi(m)}$ are the positive integers less than m and relatively prime to m, then the least residues (mod m) of $ar_1, ar_2, \ldots, ar_{\phi(m)}$ are a permutation of $r_1, r_2, \ldots, r_{\phi(m)}$. (For example, if $a = 5$ and $m = 12$, the multiples of a are 5, 25, 35, and 55, with least residues 5, 1, 11, and 7.) Since there are $\phi(m)$ elements in each set, to show that the second is a permutation of the first it suffices to show that the least residues (mod m) of $ar_1, ar_2, \ldots, ar_{\phi(m)}$ are all different and all relatively prime to m.

For the second assertion, suppose that $p \mid ar_i$ and $p \mid m$, $i = 1, 2, \ldots, \phi(m)$. Either $p \mid a$ or $p \mid r_i$. In the first case, $p \mid a$ and $p \mid m$, which is impossible because $(a, m) = 1$. In the second case, $p \mid r_i$ and $p \mid m$, which is impossible because r_i and m are relatively prime.

For the first assertion, suppose that $ar_i \equiv ar_j \pmod{m}$ for some i and j. Because $(a, m) = 1$, this implies $r_i \equiv r_j \pmod{m}$. Because r_i and r_j are least residues, this implies that $r_i = r_j$. So, $r_i \neq r_j$ implies that ar_i and ar_j are different (mod m).

Thus we know that

$$\begin{aligned} r_1 r_2 \cdots r_{\phi(m)} &\equiv (ar_1)(ar_2) \cdots (ar_{\phi(m)}) \\ &\equiv a^{\phi(m)}(r_1 r_2 \cdots r_{\phi(m)}) \pmod{m}. \end{aligned}$$

Because each of $r_1, r_2, \cdots, r_{\phi(m)}$ is relatively prime to m their product is as well, so the common factor can be cancelled to give

$$a^{\phi(m)} \equiv 1 \pmod{m},$$

which is Euler's generalization of Fermat's theorem.

So, we know that $1^4 \equiv 5^4 \equiv 7^4 \equiv 11^4 \pmod{12}$ without doing any calculation, not that calculation is needed for the first and last cases.

There was a gap of one hundred years between Fermat's theorem and Euler's generalization of it. Mathematical talent was thin on the ground in those days.

We proceed to develop a formula for calculating $\phi(n)$ that is superior to counting on our fingers the number of positive integers less than n and relatively prime to it. We first determine $\phi(n)$ for prime powers and then show that ϕ is multiplicative.

Lemma $\phi(p^n) = p^{n-1}(p-1)$.

Proof The positive integers less than or equal to p^n that are not relatively prime to p are just those that are divisible by p:

$$1 \cdot p,\ 2 \cdot p,\ \ldots,\ (p^{n-1}) \cdot p.$$

There are p^{n-1} such multiples. So the number of integers less than or equal to p that are relatively prime to p is $\phi(n) = p^n - p^{n-1} = p^{n-1}(p-1)$.

The positive integers less that 27 and prime to it are

$$1, 2, 4, 5, 7, 8, 10, 11, 13, 14, 16, 17, 19, 20, 22, 23, 25, 26$$

and there are $3^2(3-1) = 18$ of them.

Theorem *ϕ is multiplicative.*

Proof Suppose that $(m, n) = 1$. Let us write the integers from 1 to mn in a rectangular array:

$$\begin{array}{ccccc} 1 & m+1 & 2m+1 & \cdots & (n-1)m+1 \\ 2 & m+2 & 2m+2 & \cdots & (n-1)m+2 \\ & & & \vdots & \\ m & 2m & 3m & \cdots & mn \end{array}$$

We want to see how many of the integers in the array are relatively prime to mn. We do not have to look at rows whose first element r is not relatively prime to m, because if $p \mid r$ and $p \mid m$, then p will divide each integer in the row

$$r \quad m+r \quad 2m+r \quad \cdots \quad (n-1)m+r.$$

When we delete those rows we are left with $\phi(m)$ rows, those whose first elements are integers relatively prime to m:

$$s \quad m+s \quad 2m+s \quad \cdots \quad (n-1)m+s$$

with $(s, m) = 1$. Each integer in such a row is relatively prime to m. (If $p \mid m$ and $p \mid (km + s)$, then $p \mid s$, which is impossible.) Now we want to know how many integers in that row are relatively prime to n. We assert that the least residues (mod n) of the elements in the row are a permutation of

$$0, 1, \ldots, n-1.$$

To see this, because the row contains n elements, it suffices to show that they are all different (mod n). Suppose that $km + r \equiv jm + r \pmod{n}$ for some k and j. Then $km \equiv jm \pmod{n}$. Because $(m, n) = 1$, we have $k \equiv j \pmod{n}$. That is, different k and j lead to different elements in the row, showing that they are all different (mod n).

Because the list $0, 1, \ldots, n-1$ contains $\phi(n)$ elements relatively prime to n, the row with first element s does too. There are $\phi(m)$ such rows, so the total count of integers relatively prime to m and n is $\phi(m)\phi(n)$. Thus $(m, n) = 1$ implies $\phi(mn) = \phi(m)\phi(n)$ and ϕ is multiplicative.

We can now express $\phi(n)$ in terms of its prime-power decomposition.

Theorem *If* $n = p_1^{e_1} p_2^{e_2} \cdots p_k^{e_k}$ *then*

$$\begin{aligned}\phi(n) &= \phi(p_1^{e_1})\phi(p_2^{e_2})\cdots\phi(p_k^{e_k})\\ &= p_1^{e_1-1}(p_1-1)p_2^{e_2-1}(p_2-1)\cdots p_k^{e_k-1}(p_k-1)\\ &= n\left(1-\frac{1}{p_1}\right)\left(1-\frac{1}{p_2}\right)\cdots\left(1-\frac{1}{p_k}\right).\end{aligned}$$

Proof The first representation follows from the fact that ϕ is multiplicative, the second from the first because of its value at prime powers, and the third from the second by algebraic rearrangement.

For example,

$$\phi(2000) = \phi(2^4 \cdot 5^3) = \phi(2^4)\phi(5^3) = 2^3(2-1)5^2(5-1) = 8 \cdot 25 \cdot 4 = 800.$$

We will need the following result in the next chapter.

Theorem $\displaystyle\sum_{d|n} \phi(d) = n.$

For example, if $n = 15$,

$$\sum_{d|15} \phi(d) = \phi(1) + \phi(3) + \phi(5) + \phi(15) = 1 + 2 + 4 + 8 = 15.$$

It would be natural to try to use the formulas of the last theorem, but it is quicker to use a clever idea of Gauss.

Proof Take an integer k from $1, 2, \ldots, n$ and put it in class C_d if and only if $(k, n) = d$. That is, k is in C_d if and only if $(k/d, n/d) = 1$. That says that C_d contains $\phi(n/d)$ elements. Each of $1, 2, \ldots, n$ was placed in a class, so n is the total of the number of elements in each class. That is,

$$n = \sum_{d|n} \phi(n/d).$$

The sum is just $\sum_{d|n} \phi(d)$ in the reverse order, so the theorem is proved.

For example, if $n = 15$ the classes are

d	C_d	$\phi(15/d)$
1	1, 2, 4, 7, 8, 11, 13, 14	8
3	3, 6, 9, 12	4
5	5, 10	2
15	15	1

CHAPTER 14

PRIMITIVE ROOTS AND ORDERS

We know that if $(a, m) = 1$ then $a^{\phi(m)} \equiv 1 \pmod{m}$. However, there may be smaller positive integer exponents with that property. For example, $\phi(15) = 8$ so if $(a, 15) = 1$ then $a^8 \equiv 1 \pmod{15}$. But $a^4 \equiv 1 \pmod{15}$ for all a relatively prime to 15.

Definition If $(a, m) = 1$, the *order* of $a \pmod{m}$ is the smallest positive integer t such that $a^t \equiv 1 \pmod{m}$.

The table (mod 15)

a	1	2	4	7	8	11	13	14
a^2		4	1	4	4	1	4	1
a^4		1		1	1		1	

shows that the order of 1 (mod 15) is 1, the order of 4, 11, and 14 is 2, and the order of 2, 7, 8, and 13 is 4.

The order is a divisor of $\phi(m)$:

Theorem *If* $(a, m) = 1$ *and* a *has order* $t \pmod{m}$, *then* $t \mid \phi(m)$.

Proof First we note that $a^n \equiv 1 \pmod{m}$ if and only if n is a multiple of t. If $n = tq$ then

$$a^n \equiv a^{tq} \equiv (a^t)^q \equiv 1^q \equiv 1 \pmod{m}.$$

Conversely, suppose that $a^n \equiv 1 \pmod{m}$. Since t is the smallest positive integer such that $a^t \equiv 1 \pmod{m}$, we know that $n \geq t$ so we can divide n by t to get $n = tq + r$ with $0 \leq r < t$. Then

$$1 \equiv a^n \equiv a^{tq+r} \equiv (a^t)^q a^r \equiv 1^q a^r \equiv a^r \pmod{m}.$$

But there is no positive integer $r < t$ with $a^t \equiv 1 \pmod{m}$. Thus $r = 0$ and $n = tq$ is a multiple of the order t.

Because $a^{\phi(m)} \equiv 1 \pmod{m}$, we thus know that $\phi(m)$ is a multiple of t.

As an example of the application of the idea of order, we have the following result, not obvious otherwise:

Theorem *If p and q are odd primes and $q \mid (a^p - 1)$ then either $q \mid (a-1)$ or $q = 2kp + 1$ for some integer k.*

Proof Because $q \mid (a^p - 1)$ we have $a^p \equiv 1 \pmod{q}$. So the order of $a \pmod{q}$ is a divisor of p. That is, the order of a is 1 or p. If the order is 1, then $a^1 \equiv 1 \pmod{q}$ and $q \mid (a - 1)$. If the order is p, then from the last theorem $p \mid \phi(q)$, or $p \mid (q - 1)$. That is, $q - 1 = rp$ for some integer r. Since the left-hand side is even, r must be even, $r = 2k$, and $q = 2kp + 1$.

As a corollary, we have that any divisor of $2^p - 1$ has the form $2kp + 1$.

The next theorem is essentially another corollary but it is worth stating explicitly:

Theorem *If the order of $a \pmod{m}$ is t, then $a^r \equiv a^s \pmod{m}$ if and only if $r \equiv s \pmod{t}$.*

Proof Suppose that $a^r \equiv a^s \pmod{m}$. We can suppose that $r \geq s$ (if not, interchange their names). Then $a^{r-s} \equiv 1 \pmod{m}$ and we know from the first theorem that $r - s$ is a multiple of t. That is, $r \equiv s \pmod{t}$.

Conversely, suppose that $r \equiv s \pmod{t}$. Then $r = s + kt$ for some k, so

$$a^r \equiv a^{s+kt} \equiv a^s(a^t)^k \equiv a^s \pmod{m}.$$

So, because 7 has order 4 (mod 15), if $7^r \equiv 7^s \pmod{15}$ then $r \equiv s \pmod{4}$.

If no power of a smaller than $\phi(m)$ is 1 (mod m) we give a a name:

Definition If a is a least residue (mod m) and the order of a is $\phi(m)$, then a is a *primitive root* of m.

For example, 2 is a primitive root of 9 because least residues of the first $\phi(9) = 6$ powers of 2 are $2, 4, 8, 7, 5, 1 \pmod{9}$. The residues are just those that are relatively prime to 9. This happens in general, as the next theorem shows.

Theorem *If g is a primitive root of m then the least residues* (mod m) *of* $g, g^2, \ldots, g^{\phi(m)}$ *are a permutation of the* $\phi(m)$ *positive integers less than* m *and relatively prime to* m.

Proof Because the powers of g are relatively prime to m (if $p|g^k$ and $p \mid m$ then $p \mid g$ and $p \mid m$, but $(g, m) = 1$) and there are $\phi(m)$ of them, to prove that we have a permutation it is enough to show that no two powers have the same least residue (mod m). If $g^j \equiv g^k \pmod{m}$ then we know from the last theorem that $j \equiv k \pmod{\phi(m)}$. Because j and k are positive integers and are less than $\phi(m)$, we have $j = k$. So, different values of j and k give different values of g^j and g^k.

There is an analogue of the logarithm for integers mod p. The last theorem shows that if we have a primitive root, g, of a prime p (and we will shortly show that primes always have primitive roots), then each of $1, 2, \ldots, p-1$ is congruent (mod p) to a power of g.

Definition Let g be a primitive root of p. If k is the smallest positive integer such that $g^k \equiv a \pmod{p}$, then $k = \text{ind}_g a$, the *index* of a to the base g.

For example, a table of powers of the primitive root 2 of 11 is

k	1	2	3	4	5	6	7	8	9	10
2^k	2	4	8	5	10	9	7	3	6	1

and it gives a table of indices to the base 2

a	1	2	3	4	5	6	7	8	9	10
$\text{ind}_2 a$	10	1	8	2	4	9	7	3	6	5

Indices have the properties expected of a logarithm. From the table,

$$\text{ind}_2 8 \equiv \text{ind}_2 4 + \text{ind}_2 2 \pmod{10} \quad \text{and} \quad \text{ind}_2 9 \equiv 2\text{ind}_2 3 \pmod{10}.$$

In general, we have the

Theorem $\text{ind}_g ab \equiv \text{ind}_g a + \text{ind}_g b \pmod{p-1}$.

Proof Let $x = \text{ind}_g a$, $y = \text{ind}_g b$, and $z = \text{ind}_g ab$. We want to show that $z \equiv x + y \pmod{p-1}$. We know that $g^x \equiv a \pmod{p}$, $g^y \equiv b \pmod{p}$, and $g^z \equiv ab \pmod{p}$. So

$$g^z \equiv g^x g^y \equiv g^{x+y} \pmod{p}.$$

This implies $z \equiv x + y \pmod{p-1}$.

Similarly, it is possible to show that $\text{ind}_g a^r \equiv r\,\text{ind}_g a \pmod{p-1}$. Thus, given a table of indices (such things exist), it is possible to do computations as with a table of logarithms.

Not all integers have primitive roots. The theorem that shows this, which we will not prove, is

Theorem *Integers that have primitive roots are* $1, 2, 4, p^k$, *and* $2p^k$, $k \geq 1$, *for odd primes* p.

We will, however, show that all odd primes p have primitive roots, in fact $\phi(p-1)$ of them. We start with a general theorem on order:

Theorem *Suppose that* a *has order* $t \pmod{m}$. *Then* a^k *has order* t *if and only if* $(k, t) = 1$.

Proof Suppose that $(k, t) = 1$ and the order of a^k is s. Because the order of a is t, we know that $1 \equiv (a^t)^k \equiv (a^k)^t \pmod{m}$. This tells us that the order of a^k is a divisor of t. That is, $s \mid t$.

On the other hand, because the order of a^k is s we know that $1 \equiv (a^k)^s \equiv a^{ks} \pmod{m}$. This tells us that the order of a is a divisor of ks. That is, $t \mid ks$. Because $(t, k) = 1$, it follows that $t \mid s$. From $t \mid s$ and $s \mid t$ we get $s = t$.

For the converse, suppose that a and a^k have the same order t and that $(k, t) = r$. Then

$$1 \equiv a^t \equiv (a^t)^{k/r} \equiv (a^k)^{t/r} \pmod{m}.$$

Because t is the order of a^k, this says that t/r is a multiple of t. This is possible only if $r = 1$.

For example, because 2 has order 4 (mod 15) we know that 2^3 will also have order 4 but 2^2 will not (it has order 2).

This theorem will give us the number of primitive roots of a prime (after we show that it has one) because of the

Corollary *If* g *is a primitive root of* p, *then* g^k *is a primitive root of* p *if and only if* $(k, p-1) = 1$.

Proof The theorem says that if a has order $p-1 \pmod{p}$ then a^k has order $p-1$ if and only if $(k, p-1) = 1$. This is what the corollary says.

There are $\phi(p-1)$ positive integers with $(k, p-1) = 1$ and thus that many primitive roots of p.

We now proceed to show that odd primes have primitive roots.

Lemma *If f is a polynomial of degree n, then $f(x) \equiv 0 \pmod{p}$ has at most n solutions.*

This is a familiar statement for polynomial equalities $f(x) = 0$, but congruences are different from equalities. If the modulus is composite, quadratic polynomials can have more than two zeros. For example, $x^2 + x \equiv 0 \pmod{6}$ has four solutions, 0, 2, 3, and 5.

Proof Let $f(x) = a_n x^n + a_{n-1}x^{n-1} + \cdots + a_0$ have degree n, so $a_n \not\equiv 0 \pmod{p}$. We will use mathematical induction.

For $n = 1$, $a_1 x + a_0 \equiv 0 \pmod{p}$ has exactly one solution because $(a_1, p) = 1$.

Suppose that the lemma is true for polynomials of degree $n - 1$. Either $f(x) \equiv 0 \pmod{p}$ has a solution or not. If not, then the lemma is true because the congruence has 0 solutions, which is less than n. If it has a solution, then there is an integer r such that $f(r) \equiv 0 \pmod{p}$. Because $x^k - r^k$ has a factor of $x - r$ for each k, $k = 1, 2, \ldots, n$ we have

$$\begin{aligned} f(x) - f(r) &\equiv \sum_{i=0}^{n} a_i x^i - \sum_{i=0}^{n} a_i r^i \\ &\equiv \sum_{i=0}^{n} a_i (x^i - r^i) \\ &\equiv (x - r) g(x) \pmod{p}, \end{aligned}$$

where g is a polynomial of degree $n - 1$.

Suppose that s is a solution of $f(x) \equiv 0 \pmod{p}$. Then

$$0 \equiv f(s) \equiv (s - r) g(s) \pmod{p}.$$

Because p is prime, either $s - r \equiv 0 \pmod{p}$ or $g(s) \equiv 0 \pmod{p}$. By the induction assumption there are at most $n - 1$ values of s that satisfy the second congruence. Since there is one value of s that satisfies the first, there are at most n in all.

Lemma *If $d \mid (p - 1)$ then $x^d \equiv 1 \pmod{p}$ has exactly d solutions.*

Proof Because $d \mid (p - 1)$,

$$x^{p-1} - 1 = (x^d - 1)(x^{p-1-d} + x^{p-1-2d} + \cdots + 1) = (x^d - 1) g(x).$$

Fermat's Theorem tells us that $x^{p-1} \equiv 1 \pmod{p}$ has exactly $p - 1$ solutions, namely $1, 2, \ldots, p - 1$. From the last lemma we know that $g(x) \equiv$

$0 \pmod{p}$ has at most $p-1-d$ solutions. So, to get the total number, $p-1$, of solutions, $x^d \equiv 1 \pmod{p}$ must have at least d solutions. The last lemma tells us that it has at most d solutions, so it has exactly d solutions.

Theorem *Every odd prime p has $\phi(p-1)$ primitive roots.*

Proof We know that each of the least residues $1, 2, \ldots, p-1$ has an order that is a divisor of $p-1$. For each divisor t of $p-1$, let $\theta(t)$ denote the number of least residues that have order t. We know that

$$\sum_{t|(p-1)} \theta(t) = p-1.$$

We also know from Chapter 13 that

$$\sum_{t|(p-1)} \phi(t) = p-1.$$

If we can show that $\theta(t) \leq \phi(t)$ for each t, it will follow that, in order to have those equalities, we must have $\theta(t) = \phi(t)$ for all t. In particular the number of least residues with order $p-1$, which is the number of primitive roots of p, will be $\phi(p-1)$.

Choose t. If $\theta(t) = 0$, then $\theta(t) \leq \phi(t)$. If $\theta(t) \neq 0$ then there is an integer with order t. Call it a. From the last lemma we know that $x^t \equiv 1 \pmod{p}$ has exactly t solutions. The congruence is satisfied by the t integers $a, a^2, a^3, \ldots, a^t$. Because no two of them have the same least residue (mod p), their least residues give all solutions. Those that have order t are those that come from a power a^k with $(k, t) = 1$. But there are $\phi(t)$ such integers k. So in this case $\theta(t) = \phi(t)$ and hence $\theta(t) \leq \phi(t)$, which is what we need to complete the proof.

We have established also the

Corollary *If p is prime and $t \mid (p-1)$, then the number of least residues (mod p) that have order t is $\phi(t)$.*

Now that we know that p has primitive roots, we can use one of them to give another proof of part of Wilson's Theorem.

Theorem *If $p > 2$ is prime, then $(p-1)! \equiv -1 \pmod{p}$.*

Proof Let g be a primitive root of p. We know that the least residues (mod p) of $g, g^2, \ldots, g^{p-1}$ are a permutation of $1, 2, \ldots, p-1$. Multiplying, we get $1 \cdot 2 \cdots (p-1) \equiv g \cdot g^2 \cdots g^{p-1} \pmod{p}$, or

$$(p-1)! \equiv g^{1+2+\cdots+(p-1)} \equiv (g^p)^{(p-1)/2} \equiv g^{(p-1)/2} \pmod{p}.$$

We know that $g^{(p-1)/2} \equiv \pm 1 \pmod{p}$ and the plus sign is ruled out because g is a primitive root. Thus $(p-1)! \equiv -1 \pmod{p}$.

CHAPTER 15

DECIMALS

Some decimal expansions of fractions terminate, such as $\frac{3}{8} = .375$, and others, such as $\frac{2}{9} = .222\ldots$, do not. Which occurs is determined by the denominator of the fraction (if $(a, n) = 1$, the expansion of a/n terminates or fails to terminate if and only if the expansion of $1/n$ terminates or fails to terminate) so the question as to which is which is answered by the

Theorem *The decimal expansion of $1/n$ terminates if and only if $n = 2^a 5^b$ for some nonnegative integers a and b.*

Proof If $1/n = 1/(2^a 5^b)$ then multiplying the numerator and denominator by 5^{a-b} or 2^{b-a}, whichever is an integer, puts it in the form $5^{a-b}/10^a$ or $2^{b-a}/10^b$. These fractions have terminating decimal expansions, after a or b places, whichever is larger.

Conversely, suppose that the decimal expansion of $1/n$ terminates. Then

$$\begin{aligned}\frac{1}{n} &= .d_1 d_2 \ldots d_k \\ &= \frac{d_1}{10} + \frac{d_2}{10^2} + \cdots + \frac{d_k}{10^k} \\ &= \frac{10^{k-1} d_1 + 10^{k-2} d_2 + \cdots + d_k}{10^k} \\ &= \frac{m}{10^k}.\end{aligned}$$

So $mn = 10^k$ and $n \mid 10^k$. Thus only prime divisors of n are 2 and 5.

If the expansion of $1/n$ does not terminate, it is periodic and the length of the period is the order of 10 (mod n).

Theorem *If* $(n, 10) = 1$ *then the length of the period of the decimal expansion of* $1/n$ *is* r, *the smallest positive integer such that* $10^r \equiv 1 \pmod{n}$.

Proof Because $10^r \equiv 1 \pmod{n}$, we know that $kn = 10^r - 1$ for some integer k and so $1/n = k/(10^r - 1)$. Because $k < 10^r$ it is an integer with at most r digits. Let the digits of k be, from left to right, $d_{r-1}, d_{r-2}, \ldots, d_0$ (so $k = \sum_{i=0}^{r-1} d_i 10^i$). Then

$$\begin{aligned} \frac{1}{n} &= \frac{k}{10^r - 1} \\ &= \frac{d_{r-1}d_{r-2}\cdots d_0}{10^r} \cdot \frac{1}{1 - 10^{-r}} \end{aligned}$$

and, writing the last factor as a sum of a geometric series,

$$\begin{aligned} \frac{1}{n} &= (0.d_{r-1}d_{r-2}\ldots d_0)(1 + 10^{-r} + 10^{-2r} + \cdots) \\ &= \overline{d_{r-1}d_{r-2}\ldots d_0}, \end{aligned}$$

a repeating decimal with period r and digits $d_{r-1}, d_{r-2}, \ldots, d_0$.

If the expansion of $1/n$ has period length s then, starting with $\frac{1}{n} = \overline{e_{s-1}e_{s-2}\ldots e_0}$ and repeating the calculations backwards, we will arrive at $10^s \equiv 1 \pmod{n}$, which shows that s is a multiple of r. So the expansion of $1/n$ can have no period length smaller than r.

For example, $10^6 - 1 = 999999 = 7 \cdot 142857$, which is why $\frac{1}{7} = .\overline{142857}$.

The theorem requires that $(n, 10) = 1$, but dividing a fraction by 2 or 5 does not affect its period. If $n = 2^a 5^b m$ where $(m, 10) = 1$, then the length of the period of $1/n$ will be the same as that of the period of $1/m$.

Since the order of an integer (mod n) is a divisor of $\phi(n)$, the length of the period of $1/n$ is a divisor of $\phi(n)$. For some integers, such as 7, the period is as long as it can possibly be but for others, such as 13, it is not: $\phi(13) = 12$ but $1/13 = .\overline{076923}$. There is no general rule known for predicting period lengths.

What we have done for decimals could be done for integers represented in any base. For example, in the duodecimal (base 12) system, fractions have duodecimal expansions that terminate if and only if their denominators have the form $2^a 3^b$ and the length of the period of the non-terminating duodecimal expansion of $1/n$ is the order or 12 (mod n).

CHAPTER 16

QUADRATIC CONGRUENCES

After solving linear congruences $ax \equiv b \pmod{m}$ it would be natural to try to solve quadratic congruences $ax^2 + bx + c \equiv 0 \pmod{m}$. If $m = p_1^{e_1} p_2^{e_2} \cdots p_k^{e_k}$ the congruence is equivalent to the system of congruences $ax^2 + bx + c \equiv 0 \pmod{p_i^{e_i}}$, $i = 1, 2, \ldots, k$, and if we can solve those then the Chinese Remainder Theorem gives us a solution (mod m). It is the case that for $r > 1$, solutions to $ax^2 + bx + c \equiv 0 \pmod{p^r}$ can be gotten from solutions to $ax^2 + bx + c \equiv 0 \pmod{p^{r-1}}$, so we will start at the ground floor with

$$ax^2 + bx + c \equiv 0 \pmod{p},$$

p an odd prime. (If $p = 2$ and we have a quadratic congruence, $a = 1$ and we have only four possibilities for b and c. Inspection suffices to give solutions.)

We can specialize still further. Because there is an integer a' such that $aa' \equiv 1 \pmod{p}$ we can multiply by it to make the quadratic's leading coefficient 1: $x^2 + a'bx + a'c \equiv 0 \pmod{p}$. If $a'b$ is even, we can complete the square:

$$\left(x + \frac{a'b}{2}\right)^2 \equiv -a'c + \left(\frac{a'b}{2}\right)^2 \pmod{p},$$

reducing the congruence to the form $x^2 \equiv a \pmod{p}$. If $a'b$ is odd, we can add or subtract a multiple of p to get the same congruence with an even term, and then complete the square. So in either case the congruence that we need to study has the form

$$x^2 \equiv a \pmod{p}.$$

For example,

$$4x^2 + 5x + 6 \equiv 0 \pmod{11}$$

becomes, on multiplication by 3,

$$x^2 + 15x + 18 \equiv x^2 + 4x + 7 \equiv (x+2)^2 + 3 \equiv 0 \pmod{11},$$

so solving $x^2 \equiv -3 \equiv 8 \pmod{11}$ will give the solutions to the original congruence.

In what follows we will suppose that a is not a multiple of p because the congruence $x^2 \equiv 0 \pmod{p}$ presents no difficulties.

For non-zero values of a, solutions come in pairs. For example, modulo 13 we have the table of squares

x	1	2	3	4	5	6	7	8	9	10	11	12
x^2	1	4	9	3	12	10	10	12	3	9	4	1

So $x^2 \equiv a \pmod{13}$ has two solutions for $a = 1, 3, 4, 9, 10$, and 12 and no solutions for $a = 2, 5, 6, 7, 8$, and 11. This holds in general:

Theorem *If p is an odd prime and $p \nmid a$, then $x^2 \equiv a \pmod{p}$ has exactly two solutions or no solutions.*

Proof If r is a solution, so $r^2 \equiv a \pmod{p}$, another solution is $p - r$ because

$$(p-r)^2 \equiv p^2 - 2pr + r^2 \equiv r^2 \equiv a \pmod{p}.$$

It is different from r because if $p - r \equiv r \pmod{p}$ then $2r \equiv 0 \pmod{p}$; $r \not\equiv 0 \pmod{p}$ because $p \nmid a$ and $2 \not\equiv 0 \pmod{p}$ because p is odd. So, if there is one solution then there are two.

Let s be any solution. Then $r^2 \equiv s^2 \equiv a \pmod{p}$ or

$$0 \equiv r^2 - s^2 \equiv (r+s)(r-s) \pmod{p}$$

so $p \mid (r+s)(r-s)$. Thus $p \mid (r+s)$ or $p \mid (r-s)$. Because r and s are least residues and are not 0, $0 < r + s < 2p$. The only multiple of p in that range is $1 \cdot p$ so $r + s = p$ and $s = p - r$, a solution we already knew. If $p \mid (r-s)$ then, because $-p < r - s < p$, we know that $r - s = 0$ and again we get no new solution. So if $x^2 \equiv a \pmod{p}$ has a solution, it has exactly two.

The integers from 1 to $p - 1$ fall into two classes, those for which $x^2 \equiv a \pmod{p}$ has solutions and those for which it does not. We give the numbers in each class a name:

Definition If $x^2 \equiv a \pmod{p}$ has a solution, then a is a *quadratic residue* of p. If $x^2 \equiv a \pmod{p}$ has no solution, then a is a *quadratic non-residue* of p.

For example, for $p = 13$, the non-zero quadratic residues are 1, 3, 4, 9, 10, and 12 and the quadratic non-residues are 2, 5, 6, 7, 8, and 11.

We introduce some notation.

Definition The *Legendre symbol* (a/p), where p is an odd prime and $p \nmid a$ is defined by

$$(a \,/\, p) = \begin{cases} 1 & \text{if } a \text{ is a quadratic residue of } p \\ -1 & \text{if } a \text{ is a quadratic non-residue of } p. \end{cases}$$

The slash has nothing to do with division. It serves only to separate the two entries in the symbol.

For example, we have $(2 \,/\, 11) = -1$ and $(3 \,/\, 11) = 1$.

We would like to be able to determine $(a \,/\, p)$ without having to make a table of squares. An aid is Euler's Criterion:

Theorem (Euler's Criterion) *If p is an odd prime and $p \nmid a$ then*

$$(a \,/\, p) \equiv a^{(p-1)/2} \pmod{p}.$$

Proof We know that p has primitive roots. Pick one and call it g. We know that $a \equiv g^k \pmod{p}$ for some k.

If k is even then a is a quadratic residue of p because

$$a \equiv g^k \equiv (g^{k/2})^2 \pmod{p}.$$

By Fermat's theorem,

$$a^{(p-1)/2} \equiv (g^k)^{(p-1)/2} \equiv (g^{k/2})^{p-1} \equiv 1 \pmod{p}.$$

Thus for quadratic residues, $a^{(p-1)/2} \equiv 1 \equiv (p \,/\, a) \pmod{p}$.

From

$$(g^{(p-1)/2})^2 \equiv g^{p-1} \equiv 1 \pmod{p}$$

we know that $g^{(p-1)/2} \equiv \pm 1 \pmod{p}$. If g is a primitive root, $g^{(p-1)/2} \equiv 1 \pmod{p}$ is impossible because the powers of g run through the least residues $\pmod{p}$ without repetition. So $g^{(p-1)/2} \equiv -1 \pmod{p}$.

Now suppose that k is odd. Then

$$a^{(p-1)/2} \equiv (g^k)^{(p-1)/2} \equiv (g^{(p-1)/2})^k \equiv (-1)^k \equiv -1 \pmod{p}.$$

Also, a is a quadratic non-residue of p. Suppose, on the contrary, that $r^2 \equiv a \pmod{p}$. Then

$$1 \equiv r^{p-1} \equiv (r^2)^{(p-1)/2} \equiv a^{(p-1)/2} \equiv -1 \pmod{p},$$

which is impossible. Thus for quadratic non-residues,

$$a^{(p-1)/2} \equiv -1 \equiv (a \,/\, p) \pmod{p}.$$

For example, from Euler's Criterion

$$(3 \,/\, 11) \equiv 3^5 \equiv 3 \cdot 81 \equiv 3 \cdot 4 \equiv 12 \equiv 1 \pmod{11},$$

so we know, as the table of squares (mod 11) told us, that $(3 \,/\, 11) = 1$.

Euler's Criterion lets us immediately determine for which primes p it is true that $x^2 \equiv -1 \pmod{p}$ has a solution.

Corollary *If p is an odd prime, then*

$$\begin{aligned} (-1 \,/\, p) &= 1 && \text{if } p \equiv 1 \pmod{4} \\ (-1 \,/\, p) &= -1 && \text{if } p \equiv 3 \pmod{4}. \end{aligned}$$

Proof Euler's Criterion says that $(-1/p) \equiv (-1)^{(p-1)/2} \pmod{p}$.

If $p \equiv 1 \pmod{4}$, then $p = 1 + 4k$, so $(p-1)/2 = 2k$ is even and so $(-1 \,/\, p) \equiv 1 \pmod{p}$. Because the value of $(-1 \,/\, p)$ is ± 1, $(-1 \,/\, p) = 1$.

If $p \equiv 3 \pmod{4}$, then $p = 3 + 4j$, so $(p-1)/2 = 1 + 2j$ is odd and so $(-1 \,/\, p) \equiv -1 \pmod{p}$. Because the value of $(-1 \,/\, p)$ is ± 1, $(-1 \,/\, p) = -1$.

If we wanted to be picturesque, we could say that i exists modulo p (that is, -1 has a square root) for $p = 5, 13, 17, 29, \ldots$.

We will develop methods that will let us evaluate $(a \,/\, p)$ without having to calculate powers (mod p). First we need some properties of the Legendre symbol.

Theorem *The Legendre symbol has the properties*

(A) *if $a \equiv b \pmod{p}$, then $(a \,/\, p) = (b \,/\, p)$.*

(B) *if $p \nmid a$, then $(a^2 \,/\, p) = 1$.*

(C) *if $p \nmid a$ and $p \nmid b$, then $(ab \,/\, p) = (a \,/\, p)(b \,/\, p)$.*

Proof (A): Euler's Criterion tells us that

$$(a \mathbin{/} p) \equiv a^{(p-1)/2} \equiv b^{(p-1)/2} \equiv (b \mathbin{/} p) \pmod{p}.$$

Since Legendre symbols have value ± 1, the congruence implies the equality that we want.

(B): Because $x^2 \equiv a^2 \pmod{p}$ has a solution, namely a, we have that a^2 is a quadratic residue of p.

(C): Euler's Criterion tells us that

$$(ab \mathbin{/} p) \equiv (ab)^{(p-1)/2} \equiv a^{(p-1)/2} b^{(p-1)/2} \equiv (a \mathbin{/} p)(b \mathbin{/} p) \pmod{p}$$

and, once again, the congruence implies the equality.

When we have the Quadratic Reciprocity Theorem (to be found in Chapter 18) we will be able to evaluate $(a \mathbin{/} p)$ for any a.

At the beginning of the chapter we said that solutions to

$$x^2 \equiv a \pmod{p^{k+1}}$$

could be obtained from solutions to $x^2 \equiv a \pmod{p^k}$. Here is how that can be done. Suppose that r is a solution to $x^2 \equiv a \pmod{p^k}$. To solve $x^2 \equiv a \pmod{p^{k+1}}$, put $x = r + mp^k$. Then we want to find m such that $(r + mp^k)^2 \equiv a \pmod{p^{k+1}}$ or

$$r^2 + 2rmp^k + m^2 p^{2k} \equiv a \pmod{p^{k+1}}.$$

Because $r^2 \equiv a \pmod{p^k}$ we have $r^2 = a + sp^k$, so the last congruence becomes

$$sp^k + 2rmp^k \equiv 0 \pmod{p^{k+1}} \quad \text{or} \quad s + 2rm \equiv 0 \pmod{p},$$

which can be solved for m.

For example, to find a solution to $x^2 \equiv 2 \pmod{49}$, start with $x^2 \equiv 2 \pmod{7}$ with solution 3, let $x = 3 + 7m$, and solve $(3 + 7m)^2 \equiv 2 \pmod{49}$. The congruence becomes

$$9 + 42m \equiv 2 \pmod{49}, \quad \text{or} \quad -7m \pmod{49}, \quad \text{so} \quad m \equiv 1 \pmod{7},$$

giving the solution $x = 10$. That could be used to find a solution to $x^2 \equiv 2 \pmod{343}$, and so on for higher powers of 7.

CHAPTER 17

GAUSS'S LEMMA

Gauss's Lemma is needed to prove the Quadratic Reciprocity Theorem, that for odd primes p and q, $(p/q) = (q/p)$ unless $p \equiv q \equiv 3 \pmod 4$, in which case $(p/q) = -(q/p)$, but it also has other uses.

Theorem (Gauss's Lemma) *Suppose that p is an odd prime, $p \nmid a$, and that among the least residues* (mod p) *of $a, 2a, \ldots, ((p-1)/2)a$ exactly g are greater than $(p-1)/2$. Then $(a/p) = (-1)^g$.*

Proof Divide the least residues (mod p) of $a, 2a, \ldots, ((p-1)/2)a$ into two classes: $r_1, r_2, \ldots, r_k$ that are less than or equal to $(p-1)/2$ and $s_1, s_2, \ldots, s_g$ that are greater than $(p-1)/2$. Then $k + g = (p-1)/2$. By Euler's Criterion, to prove the theorem it is enough to show that $a^{(p-1)/2} \equiv (-1)^g \pmod p$.

No two of $r_1, r_2, \ldots, r_k$ are congruent (mod p). If they were we would have $k_1 a \equiv k_2 a \pmod p$ and, because $(a, p) = 1$, $k_1 \equiv k_2 \pmod p$. Because k_1 and k_2 are both in the interval $[1, (p-1)/2]$ we have $k_1 = k_2$. Similarly, no two of $s_1, s_2, \ldots, s_g$ are congruent (mod p).

Form the set

$$r_1, r_2, \ldots, r_k, p - s_1, p - s_2, \ldots, p - s_g.$$

It contains $(p-1)/2$ elements, each in the interval $[1, (p-1)/2]$. Gauss noticed, as we now show, that its elements are all different (mod p). Because the rs and ss are different among themselves, we need only show that $r_i \not\equiv p - s_j \pmod p$ for all i and j. Suppose that $r_i \equiv p - s_j \pmod p$. Because $r_i \equiv ta \pmod p$ and $s_j \equiv ua \pmod p$ for some integers t and u in the interval $[1, (p-1)/2]$, we have $ta \equiv p - ua \pmod p$ or $(t + u)a \equiv$

0 (mod p). Because a and p are relatively prime, $t + u \equiv 0 \pmod{p}$. This is impossible because $2 \le t + u \le p - 1$.

Thus the set contains $(p-1)/2$ integers, all different (mod p), all least residues (mod p), and all in the interval $\left[1, (p-1)/2\right]$. Thus its members are a permutation of $1 \cdot 2 \cdots \left((p-1)/2\right)$. So,

$$r_1 r_2 \cdots r_k (p - s_1)(p - s_2) \cdots (p - s_g) \equiv 1 \cdot 2 \cdots \left((p-1)/2\right) \pmod{p}.$$

Because $p - s_i \equiv -s_i \pmod{p}$ and there are g such terms, this becomes

$$r_1 r_2 \cdots r_k s_1 s_2 \cdots s_g (-1)^g \equiv \left((p-1)/2\right)! \pmod{p}.$$

But $r_1, r_2, \ldots, r_k, s_1, s_2, \ldots, s_g$ are, by definition, the least residues (mod m) of $a, 2a, \ldots, (p - 1/2)a$ in some order, so

$$r_1 r_2 \cdots r_k s_1 s_2 \cdots s_g \equiv a^{(p-1)/2} \left((p-1)/2\right)! \pmod{p}.$$

The last two congruences give

$$a^{(p-1)/2} (-1)^g \left((p-1)/2\right)! \equiv \left((p-1)/2\right)! \pmod{p}.$$

The common factor is relatively prime to p and may be cancelled to give

$$a^{(p-1)/2} (-1)^g \equiv 1 \pmod{p}.$$

Multiplying both sides by $(-1)^g$ gives us $a^{(p-1)/2} \equiv (-1)^g \pmod{p}$. Euler's Criterion says that $a^{(p-1)/2} \equiv (a \,/\, p) \pmod{p}$ so we have $(a \,/\, p) \equiv (-1)^g \pmod{p}$. Since both quantities are ± 1, it follows that $(a \,/\, p) = (-1)^g$.

For example, if $p = 13$ and $a = 4$, then we know that $(4 \,/\, 13) = 1$ because 4 is a square (mod 13). The first $(p-1)/2 = 6$ multiples of 4 are 4, 8, 12, 16, 20, and 24 with least residues 4, 8, 12, 3, 7, and 11. Two of them, 4 and 3, are less than 6, so $g = 2$ and $(4 \,/\, 13) = (-1)^g$.

We know that $(-1 \,/\, p) = 1$ if $p \equiv 1 \pmod{4}$ and $(-1 \,/\, p) = -1$ if $p \equiv 3 \pmod{4}$. We can use Gauss's Lemma to determine $(2 \,/\, p)$.

Theorem $(2 \,/\, p) = 1$ *if* $p \equiv 1$ *or* 7 (mod 8) *and* $(2 \,/\, p) = -1$ *if* $p \equiv 3$ *or* 5 (mod 8).

Proof To use Gauss's Lemma, we need to count how many of the least residues of $2, 4, \ldots, 2\big((p-1)/2\big)$ (mod p) are greater than $(p-1)/2$. Since the numbers in the list are already least residues (mod p), we need only see how many are greater than $(p-1)/2$. Let $2a$ be the first even integer in the list that is greater than $(p-1)/2$. In the interval $\big[2, (p-1)/2\big]$ there are $a-1$ even integers, namely $2, 4, \ldots, 2(a-1)$. So, the number of even integers from 2 to $p-1$ that are greater than $(p-1)/2$ is the total number of even integers, $(p-1)/2$, minus the number that are less than $(p-1)/2$, which is $a-1$. That is,

$$g = \frac{p-1}{2} - (a-1).$$

We consider cases.

- If $p = 8k+1$, then $2a$ is the first even integer greater than $4k$, so $2a = 4k+2$ and $a = 2k+1$. Then $g = 4k - 2k = 2k$ is even and $(2\,/\,p) = 1$.

- If $p = 8k+3$, then $2a$ is the first even integer greater than $4k+1$, so $2a = 4k+2$ and $a = 2k+1$. Then $g = 4k+1-2k = 2k+1$ is odd and $(2\,/\,p) = -1$.

- If $p = 8k+5$, then $2a$ is the first even integer greater than $4k+2$, so $2a = 4k+4$ and $a = 2k+2$. Then $g = 4k+2-(2k+1) = 2k+1$ is odd and $(2\,/\,p) = -1$.

- If $p = 8k+7$, then $2a$ is the first even integer greater than $4k+3$, so $2a = 4k+4$ and $a = 2k+2$. Then $g = 4k+3-(2k+1) = 2k+2$ is even and $(2\,/\,p) = 1$.

Though it is in general impossible to tell when 2 is a primitive root of a prime (short of trying it out) Gauss's Lemma gives as a corollary the

Theorem *If p and $4p+1$ are both primes, then 2 is a primitive root of $4p+1$.*

Proof Let $q = 4p+1$. Then $\phi(q) = 4p$ and the order of 2 (mod q) is 1, 2, 4, p, $2p$, or $4p$. We will show that the first five cases do not occur, so 2 is a primitive root of q.

By Euler's Criterion,

$$2^{(q-1)/2} \equiv 2^{2p} \equiv (2\,/\,q) \pmod{q}.$$

We know that $p \equiv 1, 3, 5,$ or $7 \pmod 8$, so $4p \equiv 4, 4, 4,$ or $4 \pmod 8$ and $q = 4p + 1 \equiv 5 \pmod 8$. We know from the last theorem that 2 is a quadratic non-residue of primes congruent to 5 (mod 8), so

$$-1 \equiv (2\,/\,q) \equiv 2^{(q-1)/2} \equiv 2^{2p} \pmod q.$$

This tells us that the order of 2 (mod q) is not $2p$. Nor can it be a divisor of $2p$, in particular 1, 2, and p. Because its order (mod q) is not 4 either—if it were, we would have $2^4 \equiv 1 \pmod q$, but because $q \geq 4 \cdot 2 + 1 = 9$ it is impossible for q to be a divisor of $2^4 - 1 = 15$—its order is $4p$ and so 2 is a primitive root of q.

CHAPTER 18

THE QUADRATIC RECIPROCITY THEOREM

We can now prove the Quadratic Reciprocity Theorem that for distinct odd primes p and q, $(p/q) = (q/p)$ unless $p \equiv q \equiv 3 \pmod 4$, in which case $(p/q) = -(q/p)$. It was first proved by Gauss, who was very fond of it, probably because of its elegance (there seems to be no reason why square roots (mod p) should be related to square roots (mod q)) and its applications to quadratic forms. He gave eight proofs and many have been found since. We give the details of Gauss's third proof.

We need a lemma:

Lemma *If p and q are different odd primes, then*

$$\sum_{k=1}^{(p-1)/2} \left[\frac{kq}{p}\right] + \sum_{k=1}^{(q-1)/2} \left[\frac{kp}{q}\right] = \frac{p-1}{2} \cdot \frac{q-1}{2}.$$

($[x]$ denotes the largest integer less than x—the integral part of x.)

Proof Let

$$S(p,q) = \sum_{k=1}^{(p-1)/2} \left[\frac{kq}{p}\right],$$

so we are trying to show that $S(p,q) + S(q,p) = (p-1)(q-1)/4$.

From Figure 18.1, we see that $S(p,q)$ is the number of lattice points—points with integer coordinates—above the x-axis and below the line $y = (q/p)x$ for $x = 1, 2, \ldots, (p-1)/2$. (There are no lattice points on the line. If there were and (a,b) was a point on the line, then $b = (q/p)a$ or

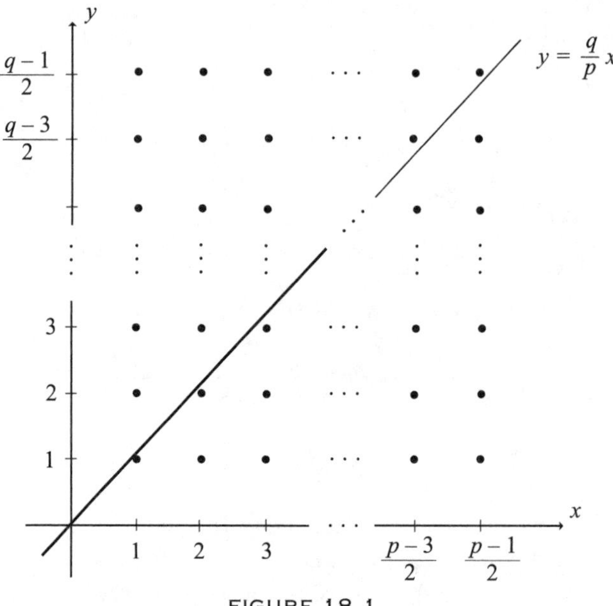

FIGURE 18.1.

$bp = qa$. This implies that $p \mid a$, but $1 \le a \le (p-1)/2$ and there are no multiples of p in that interval.)

Also, $S(q, p)$ is the number of lattice points to the left of the line and to the right of the y-axis. Thus $S(p, q) + S(q, p)$ counts each of the lattice points in the figure exactly once, giving their total number, which is

$$(p-1)/2 \cdot (q-1)/2.$$

Theorem (the Quadratic Reciprocity Theorem) *If p and q are odd primes, then*

$$(p/q)(q/p) = (-1)^{(p-1)(q-1)/4}.$$

(This is a symbolic version of the statement of the theorem given earlier: $(p-1)(q-1)/4$ is even unless $p \equiv q \equiv 3 \pmod 4$.)

Proof As in the proof of Gauss's Lemma, let us take the least residues (mod p) of $q, 2q, \ldots, \big((p-1)/2\big)q$ and divide them in two classes. Put those less than or equal to $(p-1)/2$ in one class and call them $r_1, r_2, \ldots, r_k$ and put those greater than $(p-1)/2$ in another and call them $s_1, s_2, \ldots, s_g$. In proving Gauss's Lemma we showed that the numbers

$$r_1, r_2, \ldots, r_k, p - s_1, p - s_2, \ldots, p - s_g$$

were a permutation of $1, 2, \ldots, (p-1)/2$. Let R be the sum of the rs and S the sum of the ss, so

$$\sum_{j=1}^{k} r_j + \sum_{j=1}^{g} (p - s_j) = R + gp - S.$$

Thus

$$\begin{aligned} R + gp - S &= \sum_{j=1}^{(p-1)/2} j \\ &= \frac{1}{2}\left(\frac{p-1}{2}\right)\left(\frac{p-1}{2} + 1\right) \qquad (1) \\ &= \frac{1}{8}(p-1)(p+1) = \frac{p^2-1}{8}. \end{aligned}$$

When we divide jq by p we get quotient $[jq/p]$ and remainder t_j, where t_j is the least residue of $jq \pmod{p}$:

$$\frac{jq}{p} = \left[\frac{jq}{p}\right] + \frac{t_j}{p}.$$

As j goes from 1 to $(p-1)/2$, the least residues t_j will be $r_1, r_2, \ldots, r_k$ and $s_1, s_2, \ldots, s_g$ in some order. Thus, summing,

$$\sum_{j=1}^{(p-1)/2} jq = \sum_{j=1}^{(p-1)/2} [jq/p]p + \sum_{j=1}^{(p-1)/2} t_j$$

or

$$q \sum_{j=1}^{(p-1)/2} j = p \sum_{j=1}^{(p-1)/2} [jq/p] + \sum_{j=1}^{k} r_j + \sum_{j=1}^{g} s_j$$

or

$$q(p^2-1)/8 = pS(p,q) + R + S.$$

From (1), $R = S - gp + (p^2-1)/8$. Substituting, we get

$$q(p^2-1)/8 = pS(p,q) + 2S - gp + (p^2-1)/8$$

or

$$(q-1)(p^2-1)/8 = p(S(p,q) - g) + 2S.$$

The left-hand side is even, because $(p^2-1)/8$ is an integer and $q-1$ is even. Because $2S$ is even, it follows that the remaining term, $p(S(p,q)-g)$, is even. Thus $S(p,q)-g$ is even, $S(p,q)+g$ is also, and so $(-1)^{S(p,q)+g} = (-1)^{S(p,q)}(-1)^g = 1$. From Gauss's Lemma, $(-1)^g = (q/p)$, so

$$1 = (-1)^{S(p,q)}(-1)^g \quad \text{or} \quad (-1)^{S(p,q)} = (-1)^g = (q/p).$$

We can repeat the steps with p and q interchanged—nowhere did we require p to have a property that q did not—to get

$$(-1)^{S(q,p)} = (p/q).$$

Multiplying,

$$(-1)^{S(p,q)+S(q,p)} = (p/q)(q/p),$$

and from the lemma we have

$$(-1)^{(p-1)(q-1)/4} = (p/q)(q/p),$$

which is what we want.

We can use the theorem, for example, to determine the quadratic character of 3 (mod p).

Corollary $(3\,/\,p) = 1$ *if* $p \equiv 1$ *or* 11 (mod 12) *and* $(3\,/\,p) = -1$ *if* $p \equiv 5$ *or* 7 (mod 12).

Proof The Quadratic Reciprocity Theorem tells us that $(3\,/\,p) = (p\,/\,3)$ unless $p \equiv 3 \pmod 4$, in which case $(3\,/\,p) = -(p\,/\,3)$. So, if $p = 1 + 12k$, then, using the fact that $a \equiv b \pmod 3$ implies $(a\,/\,3) = (b\,/\,3)$,

$$(3\,/\,p) = (p\,/\,3) = \big((12k+1)/3\big) = (1\,/\,3) = 1.$$

If $p = 5 + 12k$, then

$$(3\,/\,p) = (p\,/\,3) = \big((12k+5)/3\big) = (5\,/\,3) = (2\,/\,3) = -1.$$

If $p = 7 + 12k$, then

$$(3\,/\,p) = -(p\,/\,3) = -\big((12k+7)/3\big) = -(7\,/\,3) = -(1\,/\,3) = -1.$$

If $p = 11 + 12k$, then

$$(3\,/\,p) = -(p\,/\,3) = -\big((12k+11)/3\big) = -(11\,/\,3) = -(2\,/\,3) = 1.$$

If we ever need to do it, the theorem lets us evaluate $(a\,/\,p)$. For example,

$$
\begin{aligned}
&(852 / 2011) \\
&\quad = (2^2 / 2011)(3 / 2011)(71 / 2011) && \text{(because } 852 = 2^2 \cdot 3 \cdot 71\text{)} \\
&\quad = (3 / 2011)(71 / 2011) && \text{(because } (2^2 / 2011) = 1\text{)} \\
&\quad = (-1)(71 / 2011) && \text{(because } 2011 \equiv 7 \pmod{12}\text{)} \\
&\quad = (2011 / 71) && \text{(because } 2011 \equiv 71 \equiv 3 \pmod{4}\text{)} \\
&\quad = (23 / 71) && \text{(because } 2011 \equiv 23 \pmod{71}\text{)} \\
&\quad = -(71 / 23) && \text{(because } 71 \equiv 23 \equiv 3 \pmod{4}\text{)} \\
&\quad = -(2 / 23) && \text{(because } 71 \equiv 2 \pmod{23}\text{)} \\
&\quad = -1 && \text{(because } 23 \equiv 7 \pmod{8}\text{).}
\end{aligned}
$$

CHAPTER 19

The Jacobi Symbol

It is natural to ask if the Legendre symbol $(a\,/\,p)$ could be generalized to $(a\,/\,b)$ where b is not an odd prime. The Jacobi symbol does this.

Definition If a is an integer, $(a, b) = 1$, and b is an odd integer with prime factorization $b = p_1 p_2 \cdots p_k$, where the primes are not necessarily distinct, then the Jacobi symbol is defined by

$$(a\,/\,b) = \prod_{i=1}^{k} (a\,/\,p_i)$$

where the $(a\,/\,p_i)$ are Legendre symbols.

Suppose that a is a quadratic residue of b, so there is an integer r with $r^2 \equiv a \pmod{b}$. That implies that $r^2 \equiv a \pmod{p_i}$ and so $(a\,/\,p_i) = 1$ for all i. Thus $(a\,/\,b) = 1$. Unfortunately the converse does not hold as $(a\,/\,b) = 1$ does not imply that a is a quadratic residue of b. An example is

$$(2\,/\,15) = (2\,/\,3)(2\,/\,5) = (-1)(-1) = 1;$$

$x^2 \equiv 2 \pmod{15}$ has no solutions. However, if $(a\,/\,b) = -1$ then a cannot be a quadratic residue of b because at least one of the congruences $x^2 \equiv a \pmod{p_i}$ has no solution.

The Jacobi symbol is useful because it shares many properties of the Legendre symbol, including reciprocity. To get some of them we need the

Lemma *If b is an odd integer and $b = p_1 p_2 \cdots p_k$, where the primes are not necessarily distinct, then*

$$\sum_{i=1}^{k} (p_i - 1)/2 \equiv (b-1)/2 \pmod{2},$$

and hence

$$\prod_{i=1}^{k}(-1)^{(p_i-1)/2} = (-1)^{(b-1)/2}.$$

Proof Write the prime decomposition of b as

$$b = p_1 p_2 \cdots p_r p_{r+1} p_{r+2} \cdots p_k$$

where the first r primes are congruent to 3 (mod 4) and the last $k - r$ are congruent to 1 (mod 4). Thus, for $i = 1, 2, \ldots, r$, $(p_i - 1)/2 \equiv 1 \pmod 2$. For the remaining primes, $(p_i - 1)/2 \equiv 0 \pmod 2$. Thus

$$\sum_{i=1}^{k}(p_i - 1)/2 \equiv r \pmod 2.$$

We have

$$b \equiv p_1 \cdots p_r p_{r+1} \cdots p_k \equiv 3^r \cdot 1^{k-r} \equiv (-1)^r \pmod 4.$$

If r is even then $b \equiv 1 \pmod 4$ and $(b - 1)/2$ is even. If $(b - 1)/2$ is even then $b \equiv 1 \pmod 4$ and r is even. Thus $r \equiv (b - 1)/2 \pmod 2$, which gives the result.

Here are some properties of the Jacobi symbol:

Theorem *If b and c are odd integers, then*

(1) $(1 / b) = 1$.
(2) $(a^2 / b) = 1$.
(3) $(ar / b) = (a / b)(r / b)$.
(4) $(a / bc) = (a / b)(a / c)$.
(5) $(-1 / b) = 1$ *if and only if* $b \equiv 1 \pmod 4$.
(6) $(2 / b) = 1$ *if and only if* $b \equiv 1$ *or* $7 \pmod 8$.

Proof (1)–(4) follow from properties of the Legendre symbol. For (5), write b as in the proof of the lemma, with the first r primes congruent to 3 (mod 4). Because

$$(-1 / b) = \prod_{i=1}^{k}(-1 / p_i) = \prod_{i=1}^{r}(-1) \prod_{i=r+1}^{k} 1 = (-1)^r,$$

$(-1 / b) = 1$ if and only if r is even, which occurs if and only if $b \equiv 1 \pmod 4$.

(6) can be proved similarly by taking the prime decomposition of b and having the first r primes in it congruent to 3 or 5 (mod 8).

The main result is that the reciprocity theorem carries over to the Jacobi symbol:

Theorem *If a and b are odd integers, then*

$$(a\,/\,b)(b\,/\,a) = (-1)^{(a-1)(b-1)/4}.$$

Proof Let $a = p_1 p_2 \cdots p_r$ and $b = q_1 q_2 \cdots q_s$ be the prime decompositions of a and b. Then

$$\begin{aligned}
(a/b) &= \left(a\,/\,\prod_{i=1}^{s} q_i\right) \\
&= \prod_{i=1}^{s} (a\,/\,q_i) \\
&= \prod_{i=1}^{s} \left(\left(\prod_{j=1}^{r} p_j\right)\,/\,q_i\right) \\
&= \prod_{i=1}^{s} \prod_{j=1}^{r} (p_j\,/\,q_i).
\end{aligned}$$

Using the Quadratic Reciprocity Theorem that for primes p and q,

$$(p\,/\,q)(q\,/\,p) = (-1)^{(p-1)(q-1)/4},$$

we have

$$\begin{aligned}
(a\,/\,b) &= \prod_{i=1}^{s} \prod_{j=1}^{r} (p_j\,/\,q_i) \\
&= \prod_{i=1}^{s} \prod_{j=1}^{r} (-1)^{(p_j-1)(q_i-1)/4} (q_i\,/\,p_j) \\
&= (-1)^{\sum_{i=1}^{s} \sum_{j=1}^{r} (p_j-1)(q_i-1)/4} \prod_{i=1}^{s} \prod_{j=1}^{r} (q_i\,/\,p_j) \\
&= (-1)^{\sum_{i=1}^{s} (q_i-1)/2 \sum_{j=1}^{r} (p_j-1)/2} (b\,/\,a).
\end{aligned}$$

Applying the lemma to the two sums and multiplying both sides of the equation by (b / a) gives the result.

Jacobi symbols have the property that $a \equiv c \pmod{b}$ implies $(a / b) = (c / b)$, because

$$(a / b) = \prod_{i=1}^{k} (a / p_i) = \prod_{i=1}^{k} (c / p_i) = (c / b).$$

We can then evaluate Jacobi symbols without factoring the first entry, except for removing factors of 2, as in

$$\begin{aligned}(135 / 791) &= -(791 / 135) = -(116 / 135) \\ &= -(4 / 135)(29 / 135) = -(135 / 29) \\ &= -(19 / 29) = -(29 / 19) \\ &= -(-9 / 19) = -(-1 / 19) \\ &= 1.\end{aligned}$$

CHAPTER 20

PYTHAGOREAN TRIANGLES

Pythagorean triangles are right triangles whose sides have integer lengths, as in the $3, 4, 5$ and $5, 12, 13$ triangles. Their sides and hypotenuse satisfy $a^2 + b^2 = c^2$, and it is natural to ask what triples (a, b, c) satisfy that equation. It is also natural to write the equation as

$$a^2 = c^2 - b^2 = (c + b)(c - b),$$

say that the product of two numbers is a square when both factors are squares, set

$$c + b = r^2, \quad c - b = s^2, \quad a^2 = r^2 s^2,$$

solve for a, b, and c, and triumphantly think, "Done!" The idea is a good one, and works, but we have to be careful, as we are in the following development.

We want to find all solutions of $x^2 + y^2 = z^2$. If x and y are not relatively prime but have a common factor d, we may divide it out to get $(x/d)^2 + (y/d)^2 = (z/d)^2$, a Pythagorean triangle with $(x/d, y/d) = 1$. So we may derive any solution of $x^2 + y^2 = z^2$ from one in which $(x, y) = 1$ by multiplying by a suitable factor. We will call such a solution a *fundamental solution* and look for all of them. We give the final result here and then work up to it.

Theorem *All fundamental solutions of $x^2 + y^2 = z^2$ in positive integers where x, y, and z have no common factor and x is even are given by*

$$\begin{aligned} x &= 2mn \\ y &= m^2 - n^2 \\ z &= m^2 + n^2, \end{aligned}$$

where m and n are relatively positive prime integers, not both odd, and $m > n$.

First, we note that in a fundamental solution, no two of x, y, z have a common factor, because if p divides any two of them it will divide the third.

Second, we note that in a fundamental solution x and y cannot both be odd, for if they were,

$$z^2 \equiv x^2 + y^2 \equiv 1 + 1 \equiv 2 \pmod{4},$$

which is impossible because squares (mod 4) are 0 or 1. Because x and y are relatively prime they cannot both be even. Thus one is even and one is odd. We will let x denote the even member of the pair. Because even + odd = odd, z is odd.

Lemma *If $r^2 = st$ and $(s,t) = 1$, then s and t are both squares.*

Proof If s is not a square, then its prime-power decomposition contains a prime p raised to an odd power. Because $(s,t) = 1$, p does not appear in the prime-power decomposition of t. Thus p appears to an odd power in the prime-power decomposition of r^2, which is impossible. Thus s is a square. Similarly, t is a square.

Lemma *Suppose that (x, y, z) is a fundamental solution of $x^2 + y^2 = z^2$ with x even. Then there exist positive integers m and n, $m > n$, relatively prime and of opposite parity such that*

$$x = 2mn, \quad y = m^2 - n^2, \quad \textit{and} \quad z = m^2 + n^2.$$

Proof Because x is even, $x = 2r$ for some r. So $x^2 = 4r^2$ and from $x^2 = z^2 - y^2$ it follows that

$$4r^2 = (z + y)(z - y).$$

We know that z and y are both odd, so $z + y$ and $z - y$ are both even. Putting $z + y = 2s$ and $z - y = 2t$ we have $z = s + t$ and $y = s - t$, and

$$r^2 = st.$$

If we knew that $(s,t) = 1$ we could conclude by the last lemma that s and t are both squares. They are relatively prime. Suppose not: if $p \mid s$ and $p \mid t$ it follows from $z = s + t$ and $y = s - t$ that $p \mid z$ and $p \mid y$. Because z and y have no common factors, this is impossible.

So there are integers m and n, which we may assume are positive, such that $s = m^2$ and $t = n^2$. Then $x^2 = 4r^2 = 4st = 4m^2n^2$ so

$$x = 2mn.$$

Also,

$$\begin{aligned} y &= s - t = m^2 - n^2 \\ z &= s + t = m^2 + n^2. \end{aligned}$$

That $m > n$ follows from the fact that y is positive. If $p \mid m$ and $p \mid n$, then from the last equations $p \mid y$ and $p \mid z$ but this is impossible because y and z are part of a fundamental solution and thus have no common factor. If m and n are both even or both odd, then y and z would both be even, but this is impossible. So m and n have opposite parity, and we have proved what we have set out to prove.

Lemma *If $x = 2mn$, $y = m^2 - n^2$, and $z = m^2 + n^2$, then (x, y, z) is a solution of $x^2 + y^2 = z^2$. If in addition $m > n$, $(m, n) = 1$, and m and n have opposite parity then (x, y, z) is a fundamental solution.*

Proof We have

$$\begin{aligned} x^2 + y^2 &= (2mn)^2 + (m^2 - n^2)^2 \\ &= 4m^2n^2 + m^4 - 2m^2n^2 + n^4 \\ &= (m^2 + n^2)^2 \\ &= z^2, \end{aligned}$$

so we have a solution.

Let p be an odd prime such that $p \mid x$ and $p \mid y$. Because $x^2 + y^2 = z^2$, $p^2 \mid z^2$ and so, p being a prime, $p \mid z$. From $p \mid y$ and $p \mid z$ it follows that $p \mid (z + y)$ and $p \mid (z - y)$. That is, $p \mid 2m^2$ and $p \mid 2n^2$.

Because p is odd, $p \mid m^2$ and $p \mid n^2$. Because p is prime, $p \mid m$ and $p \mid n$. This contradicts $(m, n) = 1$. So the only way that x and y could fail to be relatively prime is if they both were even. Because m and n have opposite parity, $y = m^2 - n^2$ is odd. Thus $(x, y) = 1$ and we have a fundamental solution.

Proof of the theorem The last two lemmas show that if we have a fundamental solution, then it has the proper form and that if we have something of the proper form, then it leads to a fundamental solution. This is what the theorem asserts.

CHAPTER 21

$x^4 + y^4 \neq z^4$

We will show that $x^4 + y^4 = z^4$ has no solutions in positive integers by using Fermat's method of infinite descent. Fermat assumed that the equation had a solution and then showed that that implied there was a solution in smaller positive integers. That solution would then give a smaller solution in positive integers, and so on. Since an infinitely descending sequence of positive integers does not exist, the assumption is incorrect and the equation has no solutions.

It is possible that Fermat thought that his method would apply to powers other than 4 and this led him to make his famous marginal note that became known as Fermat's Last Theorem, but we cannot know for sure.

We will prove a slightly more general result:

Theorem *There are no solutions in positive integers of* $x^4 + y^4 = z^2$.

If the sum of two fourth powers cannot be a square then it cannot be a fourth power either, because a fourth power is a square (of a square).

Proof Suppose that (x, y, z) is a solution of $x^4 + y^4 = z^2$ in positive integers. We can suppose that x and y are relatively prime. If not, and $p \mid x$ and $p \mid y$, then $p^2 \mid z$, so we would have

$$\left(\frac{x}{p}\right)^4 + \left(\frac{y}{p}\right)^4 = \left(\frac{z}{p^2}\right)^2,$$

another solution. We can continue to divide by common factors until none are left.

Since $(x, y) = 1$, x and y are not both even. They are not both odd either, because then we would have $z^2 \equiv x^4 + y^4 \equiv 1 + 1 \equiv 2 \pmod 4$, which is impossible. So one is even and one is odd and we can let x denote the even number. Now we have a fundamental Pythagorean triangle—$(x^2)^2 + (y^2)^2 = z^2$—and so we know that there are integers m and n, relatively prime and not both odd such that

$$\begin{aligned} x^2 &= 2mn \\ y^2 &= m^2 - n^2 \\ z &= m^2 + n^2. \end{aligned}$$

It is not the case that n is odd, for if it were then m would be even and we would have $y^2 \equiv m^2 - n^2 \equiv 0 - 1 \equiv 3 \pmod 4$, which is impossible. So n is even and we can write $n = 2q$. Thus

$$\left(\frac{x}{2}\right)^2 = mq.$$

If m and q are relatively prime we can conclude that m and q are both squares. They are: if $p \mid m$ and $p \mid q$ then $p \mid m$ and $p \mid n$, which is impossible. Thus there are integers t and v so that

$$m = t^2 \quad \text{and} \quad q = v^2.$$

Because $(m, q) = 1$, t and v can have no common factor and are relatively prime.

In the identity

$$n^2 + (m^2 - n^2) = m^2$$

we substitute

$$n = 2q = 2v^2, \quad m^2 - n^2 = y^2, \quad \text{and} \quad m = t^2$$

to get

$$(2v^2)^2 + y^2 = (t^2)^2,$$

another Pythagorean triangle.

It is in fact a fundamental triangle. If it is not then there is a prime p such that $p \mid 2v^2$ and $p \mid y$. Thus $p \mid n$ and $p \mid y$. Because $y^2 = m^2 - n^2$, this implies that $p \mid m$ and thus that p divides both m and n, But m and n are relatively prime, so this is impossible.

Because we have a fundamental triangle, we know that there are integers M and N, relatively prime and not both odd, so that

$$\begin{aligned} 2v^2 &= 2MN \\ y &= M^2 - N^2 \\ t^2 &= M^2 + N^2. \end{aligned}$$

We have $v^2 = MN$, a square written as a product of two relatively prime integers. Hence there are integers r and s such that

$$M = r^2 \quad \text{and} \quad N = s^2.$$

Substituting, we have

$$t^2 = (r^2)^2 + (s^2)^2.$$

We have been following in the footsteps of Fermat: he no doubt started with $x^4 + y^4 = z^2$ and wanted to see where it led, perhaps hoping that, as with Pythagorean triangles, it would lead to a solution. Instead it has led to another solution of the original equation, one with

$$t^2 = m \leq m^2 < m^2 + n^2 = z \leq z^2.$$

Starting with a solution, we have found a smaller solution. This leads to the contradiction that proves the theorem.

CHAPTER 22

SUMS OF TWO SQUARES

Some positive integers can be written as a sum of two squares of integers, such as $5 = 2^2 + 1^2$, and others, such as 7, cannot. Here is how to tell them apart:

Theorem *A positive integer can be written as a sum of two squares if and only if its prime-power decomposition does not contain a prime congruent to* 3 (mod 4) *raised to an odd power.*

Thus $3^2 \cdot 5 \cdot 7^2 \cdot 13$ is a sum of two squares and $3^2 \cdot 5^2 \cdot 7^3 \cdot 13^2$ is not.

If we factor out all possible even powers of primes, an integer can be written as a square times a product of primes. So, stated in another way, the theorem says that n is a sum of two squares if and only if it has the form

$$n = N^2 p_1 p_2 \cdots p_k \quad \text{or} \quad n = 2N^2 p_1 p_2 \cdots p_k$$

where the primes are congruent to 1 (mod 4).

Though known (or guessed) earlier, the theorem was first proved by Euler in 1749.

Proof We first prove the easy half, the "only if" part, stated as the contrapositive: if n does not have the proper form then n is not the sum of two squares.

Suppose that n is divisible by p, $p \equiv 3 \pmod 4$, raised to an odd power and that $n = x^2 + y^2$. If x and y have a common factor, divide both sides of the equation by its square to get $m = u^2 + v^2$ with $(u, v) = 1$. Because the prime-power decomposition of n contains p raised to an odd power, dividing n by a square will leave at least one power of p in the prime-power decomposition of m, so m is divisible by p.

Thus u is not divisible by p because if it were then v would be also, contradicting $(u, v) = 1$. Because u and v are relatively prime, the congruence $uz \equiv v \pmod{p}$ will have a solution z. Thus

$$\begin{aligned} m &\equiv u^2 + v^2 \equiv u^2 + (uz)^2 \\ &\equiv u^2(1 + z^2) \pmod{p}. \end{aligned}$$

So $u^2(1 + z^2)$ is divisible by p. Because u is not divisible by p (if it were, from $m = u^2 + v^2$ it would follow that v is a multiple of p also, which is not the case), we have that $1 + z^2$ is divisible by p. That is, $1 + z^2 \equiv 0 \pmod{p}$ or $z^2 \equiv -1 \pmod{p}$. That says that -1 is a quadratic residue of p, which is impossible because $p \equiv 3 \pmod{4}$. Our assumption that $n = x^2 + y^2$ is therefore incorrect.

The proof in the other direction takes longer. We will need the fact that sums of two squares are closed under multiplication:

$$(a^2 + b^2)(c^2 + d^2) = (ac + bd)^2 + (ad - bc)^2 .$$

This is easy to verify by multiplication, though harder to discover in the first place. From it, if we know that primes $p \equiv 1 \pmod{4}$ are sums of two squares, then it follows (because 2 is a sum of two squares) that any integer of the form $n = N^2 p_1 p_2 \cdots p_k$ or $n = 2N^2 p_1 p_2 \cdots p_k$ is a sum of two squares.

For example, we can build up the representation of 765 as a sum of two squares from the representations of its factors:

$$\begin{aligned} 765 &= 3^2 \cdot 5 \cdot 17 \\ &= 3^2(2^2 + 1^2)(4^2 + 1^2) \\ &= 3^2\big((2 \cdot 4 + 1 \cdot 1)^2 + (2 \cdot 1 - 1 \cdot 4)^2\big) \\ &= 3^2(9^2 + 2^2) = 27^2 + 6^2. \end{aligned}$$

We prove that $p \equiv 1 \pmod{4}$ is a sum of two squares by using a method of descent. We get a multiple of p as a sum of two squares and from that we represent a smaller multiple of p as a sum of two squares. The multipliers of p will, when we repeat the process, form a decreasing sequence of positive integers that must eventually reach 1.

Because -1 is a quadratic residue of p, we know that there exists u such that $u^2 \equiv -1 \pmod{p}$. That is, $u^2 + 1$ is a multiple of p, or $u^2 + 1 = kp$ for some k. We have thus found x and y such that $x^2 + y^2 = kp$, in fact with $y = 1$, but we do not need that.

We now start to descend. Let r and s be the numerically least residues of x and y (mod k). That is, let r and s be such that

$$r \equiv x \pmod{k}, \quad s \equiv y \pmod{k}, \quad \text{with } |r| \leq k/2,\ |s| \leq k/2.$$

Because $r^2 + s^2 \equiv x^2 + y^2 \pmod{k}$ and $x^2 + y^2$ is a multiple of k, so is $r^2 + s^2$. That is, $r^2 + s^2 = km$ for some m. So we have

$$(r^2 + s^2)(x^2 + y^2) = (km)(kp) = mk^2 p \tag{1}$$

and

$$(r^2 + s^2)(x^2 + y^2) = (rx + sy)^2 + (ry - sx)^2. \tag{2}$$

The terms on the right of (2) are multiples of k^2:

$$\begin{aligned} rx + sy &\equiv r^2 + s^2 \equiv 0 \pmod{k} \\ ry - sx &\equiv rs - sr \equiv 0 \pmod{k}. \end{aligned}$$

So, dividing (2) by k^2 and using (1) gives

$$mp = \left(\frac{rx + sy}{k}\right)^2 + \left(\frac{ry - sx}{k}\right)^2,$$

another multiple of p that is a sum of two squares.

It is in fact a smaller multiple of p than the one we first had. Because r and s are numerically least residues,

$$r^2 + s^2 \leq (k/2)^2 + (k/2)^2 = k^2/2.$$

We know that $r^2 + s^2 = km$, so $km \leq k^2/2$ and thus $m \leq k/2 < k$.

To complete the proof, we need only be sure that we have not descended too far, and show that $m \geq 1$. If $m = 0$, then $r^2 + s^2 = 0$ so $r = s = 0$, and thus $k|x$ and $k|y$. Thus $k|p$ and so $k = 1$ or $k = p$. If $k = 1$, then we had $x^2 + y^2 = p$ to start with and did not need to do a descent. If $k = p$, then we have $u^2 + 1 = p^2$, which is impossible.

For example, if $p = 13$, we have $8^2 + 1^2 = 5 \cdot 13$. With $k = 5$, we have $r = -2$ and $s = 1$, and

$$325 = \left((-2)^2 + 1^2\right)\left(8^2 + 1^2\right) = (-15)^2 + (-10)^2.$$

This gives $13 = 3^2 + 2^2$ and no further descent is needed.

CHAPTER 23

SUMS OF THREE SQUARES

Theorem *A positive integer is a sum of three integer squares if and only if it is not of the form* $4^a(8k+7)$.

Proof ("only if") We want to show that any integer of the form $n = 4^a(8k+7)$ is not a sum of three squares. Suppose that

$$n = 4^a(8k+7) = x^2 + y^2 + z^2$$

for some integers x, y, and z with $a > 1$. Then $x^2 + y^2 + z^2 \equiv 0 \pmod 4$. Since squares (mod 4) are either 0 or 1, this is possible only if $x^2 \equiv y^2 \equiv z^2 \equiv 0 \pmod 4$. That is, x, y and z are all even, so

$$\frac{n}{4} = \left(\frac{x}{2}\right)^2 + \left(\frac{y}{2}\right)^2 + \left(\frac{z}{2}\right)^2 = 4^{a-1}(8k+7)$$

is also a sum of three squares. Continuing the descent, we will eventually arrive at a representation of $8k+7$ as a sum of three squares. That is, we will have a sum of three squares that is congruent to 7 (mod 8). However, squares (mod 8) are 0, 1, or 4 because

$$0^2 \equiv 4^2 \equiv 0,\ 1^2 \equiv 3^2 \equiv 5^2 \equiv 7^2 \equiv 1,\ 2^2 \equiv 6^2 \equiv 4 \pmod 8$$

and no sum of three of 0, 1, and 4 can be 7 (mod 8).

The proof of the converse, that every integer not of that form is a sum of three squares, is harder. The proof given by Gauss in 1801 involves ternary quadratic forms, a topic not often seen in elementary number theory texts.

CHAPTER 24

Sums of Four Squares

Euler tried for years to show that every positive integer is a sum of four squares, but the first proof was given by Lagrange in 1770. It uses Euler's identity that shows that the product of two sums of four squares is a sum of four squares:

Lemma

$$\begin{aligned}(a^2+b^2+c^2+d^2)&(r^2+s^2+t^2+u^2)\\ &= (ar+bs+ct+du)^2+(as-br+cu-dt)^2\\ &\quad +(at-bu-cr+ds)^2+(au+bt-cs-dr)^2.\end{aligned}$$

Proof Multiply it out.

The identity becomes obvious when viewed as a statement about quaternions, but quaternions were not discovered until more than a century after Euler.

Theorem *Every positive integer is a sum of four integer squares.*

Proof If we can show that every prime is a sum of four squares, the lemma lets us conclude that every product of primes, and hence every positive integer, is a sum of four squares. Because $2 = 1^2+1^2+0^2+0^2$, we need to show only that every odd prime is a sum of four squares.

We proceed as for sums of two squares: get a multiple of p as a sum of four squares and use that representation to get a smaller multiple of p as a sum of four squares. Repeating the descent sooner or later gives p as a sum of four squares.

What gets us started is the result that if p is an odd prime, then

$$1 + x^2 + y^2 \equiv 0 \pmod{p}$$

has a solution with $0 \le x < p/2$ and $0 \le y < p/2$.

To prove this, we look, modulo p, at the numbers in the two sets

$$S_1 = \left\{ 0^2, 1^2, 2^2, \ldots, \left(\frac{p-1}{2}\right)^2 \right\}$$

and

$$S_2 = \left\{ -1 - 0^2,\ -1 - 1^2,\ -1 - 2^2, \ldots, -1 - \left(\frac{p-1}{2}\right)^2 \right\}.$$

No two numbers in S_1 are congruent (mod p). Suppose that two were. If $a^2 \equiv b^2 \pmod{p}$, then $(a+b)(a-b) \equiv 0 \pmod{p}$. If a and b are different and between 0 and $(p-1)/2$, then $0 < a + b < p - 1$ and $-(p-1)/2 \le a - b \le (p-1)/2$ so neither $a + b$ nor $a - b$ is divisible by p. It can be shown similarly that no two numbers in S_2 are congruent (mod p).

S_1 and S_2 together contain

$$\left(\frac{p-1}{2} + 1\right) + \left(\frac{p-1}{2} + 1\right) = p + 1$$

integers. There are only p residue classes (mod p), so some two of them must fall in the same class and be congruent (mod p). Because the two integers cannot both be from the same set, some member of S_1 is congruent to some member of S_2. That is, there exist x and y such that

$$x^2 \equiv -1 - y^2 \pmod{p}$$

with $0 \le x \le (p-1)/2$ and $0 \le y \le (p-1)/2$.

For example, if $p = 13$, we have

$$S_1 = \{0, 1, 4, 9, 16, 25, 36\} \quad \text{and} \quad S_2 = \{-1, -2, -5, -10, -17, -26, -37\}$$

whose entries (mod 13) are, respectively,

$$\{0, 1, 4, 9, 3, 12, 10\} \quad \text{and} \quad \{12, 11, 8, 3, 9, 0, 2\}.$$

We thus have four possibilities for x, namely 0, 3, 9, and 12.

Because $x^2 + y^2 + 1 \equiv 0 \pmod{p}$ we have found a multiple of p that is a sum of four squares:

$$kp = x^2 + y^2 + 1^2 + 0^2.$$

Furthermore, $k < p$ because

$$x^2 + y^2 + 1 < p^2/4 + p^2/4 + 1 < p^2.$$

We now have what we need for the descent. We will show that if k and p are odd, $1 < k < p$, and

$$kp = x^2 + y^2 + z^2 + w^2$$

then there is a positive integer m with $m < k$ such that

$$mp = r^2 + s^2 + t^2 + u^2$$

for some integers r, s, t, u.

The proof uses the same idea as was used in the two-squares theorem but, since there are more variables, it is longer.

We may assume that k is odd. If the k that we got from our sets S_1 and S_2 were even, or if the descent produces an even k, then we know that x, y, z, and w are all even, all odd, or two are odd and two are even. We can then rearrange the terms, if necessary, so that x and y have the same parity and the same is true for z and w. Then we have

$$\frac{kp}{2} = \left(\frac{x-y}{2}\right)^2 + \left(\frac{x+y}{2}\right)^2 + \left(\frac{z-w}{2}\right)^2 + \left(\frac{z+w}{2}\right)^2.$$

If $k/2$ is even we can repeat the process to get $(k/4)p$ as a sum of four squares. Eventually we will have an odd multiple of p written as a sum of four squares.

Choose A, B, C, and D such that

$$A \equiv x,\ B \equiv y,\ C \equiv z,\ D \equiv w \pmod{k}$$

and so that each lies strictly between $-k/2$ and $k/2$. We can do this because k is odd. It follows that

$$A^2 + B^2 + C^2 + D^2 \equiv x^2 + y^2 + z^2 + w^2 \pmod{k}$$

so

$$A^2 + B^2 + C^2 + D^2 = km$$

for some m. Because

$$A^2 + B^2 + C^2 + D^2 < k^2/4 + k^2/4 + k^2/4 + k^2/4 = k^2$$

we know that $0 \le m < k$.

If $m = 0$, then $A = B = C = D = 0$ and hence x, y, z, and w are all divisible by k. Thus $kp = x^2 + y^2 + z^2 + w^2$ is divisible by k^2. That means that p is divisible by k but that is impossible because $1 < k < p$. So we know that $m \neq 0$ and $0 < m < k$.

We will now construct a sum of four squares equal to mp, which will complete the proof. We know that

$$k^2mp = (kp)(km) = (x^2 + y^2 + z^2 + w^2)(A^2 + B^2 + C^2 + D^2)$$

so using Euler's identity we have

$$\begin{aligned} k^2mp = {} & (xA + yB + zC + wD)^2 + (xB - yA + zD - wC)^2 \\ & + (xC - yD - zA + wB)^2 + (xD + yC - zB - wA)^2. \end{aligned}$$

The terms in parentheses are each divisible by k:

$$\begin{aligned} xA + yB + zC + wD &\equiv x^2 + y^2 + z^2 + w^2 &\equiv 0 \pmod{k} \\ xB - yA + zD - wC &\equiv xy - yx + zw - wz &\equiv 0 \pmod{k} \\ xC - yD - zA + wB &\equiv xz - yw - zx + wy &\equiv 0 \pmod{k} \\ xD + yC - zB - wA &\equiv xw + yz - zy - wx &\equiv 0 \pmod{k}. \end{aligned}$$

So, if we put

$$\begin{aligned} r &= (xA + yB + zC + wD)/k \\ s &= (xB - yA + zD - wC)/k \\ t &= (xC - yD - zA + wB)/k \\ u &= (xD + yC - zB - wA)/k \end{aligned}$$

we have

$$r^2 + s^2 + t^2 + u^2 = (k^2mp)/k^2 = mp.$$

Because $m < k$ we are done.

CHAPTER 25

WARING'S PROBLEM

In 1770, Waring asserted that every positive integer is a sum of four squares, nine cubes, nineteen fourth powers, and so on.

Let $g(k)$ denote the smallest number of kth powers needed to represent every positive integer as a sum.

Theorem $g(k) \geq 2^k + \left[(3/2)^k\right] - 2.$

Proof Let $n = 2^k\left[(3/2)^k\right] - 1$. We show that to write n as a sum of kth powers takes at least $2^k + \left[(3/2)^k\right] - 2$ terms.

Because $n \leq 2^k(3/2)^k - 1 = 3^k - 1$, writing n as a sum of kth powers will use only 2^ks and 1^ks. To use as few terms as possible, we want to use as many 2^ks as possible. This number will be q where $n = 2^k q + r$, $0 \leq r < 2^k$. There will be r 1^ks so the total numbers of terms will be $q + r$. Because

$$n = 2^k\left(\left[(3/2)^k\right] - 1\right) + (2^k - 1)$$

we have $q = \left[(3/2)^k\right] - 1$ and $r = 2^k - 1$ from which the result follows.

For example, $79 = 4 \cdot 2^4 + 15 \cdot 1^4$ requires nineteen fourth powers.

The value of $g(k)$ is known for every k. There is a formula that says that

$$g(k) = 2^k + \left[(3/2)^k\right] - 2$$

except in some exceptional cases. It is suspected that the exceptional cases do not in fact occur. If they do not, the last equation holds universally. In particular,

$$g(4) = 2^4 + \left[(3/2)^4\right] - 2 = 16 + [5.0625] - 2 = 16 + 5 - 2 = 19.$$

Numbers such as 79 are essentially accidents, and a more fundamental quantity than $g(k)$ is $G(k)$, the number of kth powers needed to represent every sufficiently large integer. It is known that $G(4) = 16$. Unfortunately, besides $G(2) = 4$ no other value of G is known, though there are upper bounds. The conjecture is that $G(3) = 6$, but a proof seems a long way off.

CHAPTER 26

PELL'S EQUATION

Pell's equation antedates John Pell (1611–1685) by several centuries and Pell never solved it, but the vagaries of mathematical nomenclature are such that his name is irrevocably linked to it.

The problem is to solve $x^2 - Ny^2 = 1$ in positive integers, N not a perfect square. (If N is a perfect square, the left-hand side can be factored and the equation is no longer interesting.) The equation arises in many places. As we will show, if we have the smallest solution in positive integers, we have them all. For example, the smallest solution of $x^2 - 2y^2 = 1$ is $(3, 2)$ and powers of $3+2\sqrt{2}$ provide the others: $(3+2\sqrt{2})^2 = 17+12\sqrt{2}$ gives $(17, 12)$, $(3+2\sqrt{2})^3$ gives another solution, and so on.

The smallest solution in positive integers can be gotten from the continued fraction expansion of $\sqrt{N}$. After we have it, Pell's equation holds no terrors.

Definition We will say that the irrational number $\alpha = r + s\sqrt{N}$ *gives a solution of* $x^2 - Ny^2 = 1$, or *solves* $x^2 - Ny^2 = 1$ for short, if $r^2 - Ns^2 = 1$.

Lemma *If* $N > 0$ *is not a square and* $x + y\sqrt{N} = u + v\sqrt{N}$, *then* $x = u$ *and* $y = v$.

Proof If $x + y\sqrt{N} = u + v\sqrt{N}$ with $y \neq v$ then

$$\sqrt{N} = \frac{x-u}{v-y}$$

is a rational number, which is impossible. Thus $y = v$ and hence $x = u$.

Lemma *The set of numbers of the form* $x^2 - Ny^2$ *is closed under multiplication:*

$$(x^2 - Ny^2)(u^2 - Nv^2) = (xu + Nyv)^2 - N(xv + yu)^2.$$

Proof Multiplying out both sides establishes the identity.

Lemma *If* α *gives a solution of* $x^2 - Ny^2 = 1$, *then so does* $1/\alpha$.

Proof If $\alpha = r + s\sqrt{N}$ solves $x^2 - Ny^2 = 1$ then $r^2 - Ns^2 = 1$ and

$$\frac{1}{\alpha} = \frac{1}{r + s\sqrt{N}} \cdot \frac{r - s\sqrt{N}}{r - s\sqrt{N}} = \frac{r - s\sqrt{N}}{r^2 - Ns^2} = r - s\sqrt{N}.$$

Because $r^2 - N(-s)^2 = 1$, $1/\alpha$ solves $x^2 - Ny^2 = 1$.

Lemma *If* α *and* β *solve* $x^2 - Ny^2 = 1$ *so does* $\alpha\beta$, *and* α^k, $k = \pm 1, \pm 2, \ldots$.

Proof Suppose that $\alpha = r + s\sqrt{N}$ and $\beta = t + u\sqrt{N}$ both solve $x^2 - Ny^2 = 1$. Then

$$\alpha\beta = \left(r + s\sqrt{N}\right)\left(t + u\sqrt{N}\right) = (rt + Nsu) + (ru + st)\sqrt{N}.$$

From the lemma on closure under multiplication,

$$(rt + Nsu)^2 - N(ru + st)^2 = (r^2 - Ns^2)(t^2 - Nu^2) = 1 \cdot 1 = 1,$$

which shows that $\alpha\beta$ solves $x^2 - Ny^2 = 1$.

Letting $\beta = \alpha$ we see that α^2 solves $x^2 - Ny^2 = 1$, and hence so does $(\alpha^2)\alpha = \alpha^3$, $(\alpha^3)\alpha = \alpha^4$, and so on. Thus α^k solves $x^2 - Ny^2 = 1$ for $k = 1, 2, \ldots$. A previous lemma says that if α solves the equation then so does the reciprocal of α, so α^k for $k = -1, -2, \ldots$ solves the equation as well.

This tells us that if we have the smallest solution then we have infinitely many—those that come from powers of the α associated with it. After one more lemma we will show that any solution is one of them, so once we find the smallest solution in positive integers, we have found them all.

Lemma *Suppose that* r, s, t, *and* u *are nonnegative and that* $\alpha = r + s\sqrt{N}$ *and* $\beta = t + u\sqrt{N}$ *solve* $x^2 - Ny^2 = 1$. *Then* $\alpha < \beta$ *if and only if* $r < t$.

Proof Suppose that $r < t$. Then $r^2 < t^2$ and because $r^2 = 1 + Ns^2$ and $t^2 = 1 + Nu^2$ we know that $Ns^2 < Nu^2$. It follows that $s < u$ and thus that $\alpha < \beta$.

Conversely, suppose that $\alpha < \beta$. If $r \geq t$ then $r^2 \geq t^2$, so $1 + Ns^2 \geq 1 + Nu^2$, from which we get $s \geq v$, and thus $\alpha \geq \beta$, a contradiction.

Now we are able to describe all the solutions of $x^2 - Ny^2 = 1$. The solution $x = 1$, $y = 0$ is trivial. As we will see in the chapter on continued fractions (Chapter 27), non-trivial solutions exist. In the set of all real numbers that give a solution to the equation let $\theta > 1$ be the smallest. From the last lemma, numbers $\alpha = r + s\sqrt{N}$ that give solutions can be ordered by r and a non-empty set of positive integers has a least element. We will call θ the *generator* for $x^2 - Ny^2 = 1$.

Theorem *Let θ be the generator for $x^2 - Ny^2 = 1$. Then all non-trivial solutions with x and y positive are given by θ^k, $k = 1, 2, \ldots$.*

Proof We have already seen that powers of θ give solutions of $x^2 - Ny^2 = 1$. We now show that there are no others.

Let $x = r$, $y = s$ be any non-trivial solution of $x^2 - Ny^2 = 1$ with r and s positive and put $\alpha = r + s\sqrt{N}$. We know that $\alpha \geq \theta$ by the definition of generator, so there is a positive integer k such that

$$\theta^k \leq \alpha < \theta^{k+1} \quad \text{or} \quad 1 \leq \theta^{-k}\alpha < \theta.$$

We know that θ^{-k} and α give solutions of $x^2 - Ny^2 = 1$ and so $\theta^{-k}\alpha$ does also. But it is less than θ, which gives the smallest non-trivial solution. Thus it gives the trivial solution $x = 1$, $y = 0$, so $1 = \theta^{-k}\alpha$ and $\alpha = \theta^k$. So powers of θ give all the solutions.

For example, once we have found the smallest solution of $x^2 - 5y^2 = 1$, namely $x = 9$ and $y = 4$, we know that all solutions are given by the powers $\left(9 + 4\sqrt{5}\right)^k$, $k = 1, 2, \ldots$. The next smallest, $x = 161$, $y = 72$, comes from $\left(9 + 4\sqrt{5}\right)^2 = 81 + 2 \cdot 9 \cdot 4\sqrt{5} + 16 \cdot 5$.

In the chapter on continued fractions we will see how the continued fraction expansion of $\sqrt{N}$ can be used to show that θ exists and, moreover, to find it.

CHAPTER 27

Continued Fractions

We can write a rational number, for example 23/17, as a *continued fraction*:

$$\frac{23}{17} = 1 + \frac{6}{17} = 1 + \frac{1}{17/6} = 1 + \frac{1}{2 + 5/6}$$
$$= 1 + \cfrac{1}{2 + \cfrac{1}{6/5}}$$
$$= 1 + \cfrac{1}{2 + \cfrac{1}{1 + \cfrac{1}{5}}}.$$

Any rational number can be put in this form, which we will write as $[a_0, a_1, \ldots, a_k]$, so $23/17 = [1, 2, 1, 5]$. Any finite continued fraction represents a rational number, which can be determined by evaluating the fraction from the bottom up. The representation is unique, except that

$$[a_0, a_1, \ldots, a_k] = [a_0, a_1, \ldots, a_k - 1, 1].$$

The process used to generate the continued fraction,

$$\begin{aligned} a &= [a] + r_1,\ a_0 = [a], \\ 1/r_1 &= [1/r_1] + r_2,\ a_1 = [1/r_1], \\ 1/r_2 &= [1/r_2] + r_3,\ a_2 = [1/r_2], \\ &\vdots \end{aligned}$$

can be applied to irrational numbers as well to give infinite continued fraction representations. (If the continued fraction were finite, it would represent a rational number.) For example,

$$\begin{aligned}\sqrt{3} &= 1 + (\sqrt{3} - 1),\\ \frac{1}{\sqrt{3}-1} &= \frac{\sqrt{3}+1}{2} = 1 + \frac{\sqrt{3}-1}{2},\\ \frac{2}{\sqrt{3}-1} &= \sqrt{3} + 1 = 2 + (\sqrt{3} - 1).\end{aligned}$$

Because we have the same remainder as in the first step, the steps repeat and we have $\sqrt{3} = [1, 1, 2, 1, 2, 1, 2, \ldots] = [1, \overline{1, 2}]$, a periodic continued fraction.

What happens for $\sqrt{3}$ occurs in general—square roots have periodic continued fraction representations. We have the

Theorem *If N is not a perfect square, then $\sqrt{N} = [a_0, \overline{a_1, a_2, \ldots, a_k}]$.*

We will not prove it. The reason it is true is like the reason why reciprocals of integers have periodic decimal expansions—there are only finitely many remainders and sooner or later they must repeat.

If we truncate a continued fraction, we get a convergent.

Definition The kth *convergent* of $[a_0, a_1, a_2, \ldots]$ is

$$c_k = \frac{r_k}{s_k} = [a_0, a_1, \ldots, a_k].$$

As could be expected from the name, the sequence of convergents converges to the value of the continued fraction. Their values can be calculated using recursion relations:

Theorem *For $k \geq 1$,*

$$\begin{aligned}r_{k+1} &= a_{k+1} r_k + r_{k-1}\\ s_{k+1} &= a_{k+1} s_k + s_{k-1}.\end{aligned}$$

The proof is by mathematical induction and so will be omitted. To start the recursions we can take $r_{-1} = 1, s_{-1} = 0$, and $r_0 = a_0, s_0 = 1$.

For example, the convergents to $23/17$, findable from its expansion or from the recursion relations, are $1/1$, $3/2$, $4/3$, and $23/17$.

Also provable by induction is the

Theorem *For $k \geq 1$,*

$$r_k s_{k-1} - r_{k-1} s_k = (-1)^{k+1}.$$

This theorem allows us to solve $ax + by = 1$ for relatively prime integers a and b.

Corollary *If $a/b = [a_0, a_1, \ldots, a_k]$ then $ax + by = 1$ has solution*

$$x = (-1)^{k+1} s_{k-1}, \quad y = (-1)^k r_{k-1}.$$

Proof The last convergent of a finite continued fraction is the rational number that it represents. Thus $r_k = a$, $s_k = b$, and from the last theorem, $as_{k-1} - r_{k-1}b = (-1)^{k+1}$, which gives the result.

For example, for $23x + 17y = 1$ we have, using the recursion relations,

n	-1	0	1	2	3
a_n		1	2	1	5
r_n	1	1	3	4	23
s_n	0	1	2	3	17

and $4 \cdot 17 - 3 \cdot 23 = -1$, so $x = 3$ and $y = -4$ is a solution.

We can show that the sequence of convergents does in fact converge:

Theorem *For $k \geq 1$,*

$$|c_k - c_{k+1}| = \left| \frac{r_k}{s_k} - \frac{r_{k+1}}{s_{k+1}} \right| < \frac{1}{s_k^2}.$$

Proof From the last theorem,

$$\begin{aligned} |c_k - c_{k+1}| &= \left| \frac{r_k}{s_k} - \frac{r_{k+1}}{s_{k+1}} \right| \\ &= \left| \frac{r_k s_{k+1} - r_{k+1} s_k}{s_k s_{k+1}} \right| \\ &= \left| \frac{(-1)^{k+1}}{s_k s_{k+1}} \right| \\ &= \frac{1}{s_k s_{k+1}} < \frac{1}{s_k^2}. \end{aligned}$$

It is possible to show, by induction if nothing else, the

Theorem *If $\alpha = [a_0, a_1, \ldots,]$ with convergents $c_k = r_k/s_k$, then for $k > 2$,*

$$c_k - c_{k-1} = \frac{(-1)^{k+1}}{s_k s_{k-1}} \quad \textit{and} \quad c_k - c_{k-2} = \frac{(-1)^k a_k}{s_k s_{k-2}}.$$

The second equation shows that

$$c_0 < c_2 < c_4 < \cdots \quad \text{and} \quad c_1 > c_3 > c_5 > \cdots.$$

Using the first we see that

$$c_0 < c_2 < c_4 < \cdots < c_5 < c_3 < c_1.$$

Thus the sequences of even and odd convergents approach limits and, because $s_k \to \infty$ as $k \to \infty$, the limits are the same.

It can be shown that convergents are the best rational approximations to the number to which they converge in the sense that if r_k/s_k is a convergent to α then no rational number with a denominator smaller than s_k is closer to α than r_k/s_k.

Much more could be said about continued fractions, not only those of our simple form but generalizations as

$$a_0 + \cfrac{b_0}{a_1 + \cfrac{b_1}{a_2 + \cfrac{b_2}{a_3 + \cdots}}}$$

where the as and bs could be real or complex numbers, or even variables. Books have been written about continued fractions, and lifetimes could be devoted to them.

We state without proof the result that lets us find the smallest positive integer solution of the Pell equation $x^2 - Ny^2 = 1$.

Theorem *If N is not a perfect square then $x^2 - Ny^2 = 1$ has a non-trivial solution. Let the continued fraction expansion of $\sqrt{N}$ be $[a_0, \overline{a_1, a_2, \ldots, a_k}]$. If k, the length of the period, is even, the smallest solution of $x^2 - Ny^2 = 1$ is given by the numerator and denominator of the convergent just before the end of the first period. If the length of the period is odd, the solution is given by the convergent just before the end of the second period.*

That is, for even period length the solution is

$$x = r_{k-1}, \quad y = s_{k-1},$$

where

$$\frac{r_{k-1}}{s_{k-1}} = c_{k-1} = [a_0, a_1, \ldots, a_{k-1}]$$

and for odd period length it is

$$x = r_{2k-1}, \quad y = s_{2k-1},$$

where

$$\frac{r_{2k-1}}{s_{2k-1}} = c_{2k-1} = [a_0, a_1, \ldots, a_k, a_1, a_2, \ldots, a_{k-1}].$$

For example, $\sqrt{3} = [1, \overline{1, 2}]$ so the solution $(2, 1)$ to $x^2 - 3y^2 = 1$ is given by $[1, 1] = 2/1$. For $x^2 - 29y^2 = 1$ we have an odd period length, $\sqrt{29} = [5, \overline{2, 1, 1, 2, 10}]$, so the smallest solution $(9801, 1820)$ comes from $[5, 2, 1, 1, 2, 10, 2, 1, 1, 2]$:

n	-1	0	1	2	3	4	5	6	7	8	9
a_n		5	2	1	1	2	10	2	1	1	2
r_n	1	5	11	16	27	70	727	1524	2251	3775	9801
s_n	0	1	2	3	5	13	135	283	418	701	1820

The convergent just before the end of the first period, $r_4/s_4 = 70/13$, gives a solution to $x^2 - 29y^2 = -1$ because $4900 - 29 \cdot 169 = -1$. The other convergents, $r_n/s_n, n < k$, provide the smallest solutions to $x^2 - 29y^2 = d$ for those d such that $r_n^2 - 29s_n^2 = d$. This holds not just for 29 but in general.

CHAPTER 28

MULTIGRADES

Multigrades consist of integers whose sums are the same for more than one power, such as

$$1^k + 5^k + 8^k + 12^k = 2^k + 3^k + 10^k + 11^k$$

for $k = 1, 2, 3$. We can write this as $[1, 5, 8, 12] =^3 [2, 3, 10, 11]$.

It can be shown that to have agreement of the first k powers, a multigrade must have at least $k + 1$ elements on each side.

Multigrades are mostly of recreational interest. Two theorems that can be verified by algebra are:

Theorem *If*

$$[a_1, a_2, \dots, a_r] =^k [b_1, b_2, , \dots, b_r]$$

then

$$[a_1 + c, a_2 + c, \dots, a_r + c] =^k [b_1 + c, b_2 + c, \dots, b_r + c].$$

This allows multigrades to be normalized with smallest entry 0 or 1.

Theorem *If*

$$[a_1, a_2, \dots, a_r] =^k [b_1, b_2, \dots, b_r]$$

then

$$[a_1, a_2, \dots, a_r, b_1 + c, b_2 + c, \dots, b_r + c]$$
$$=^{k+1} [b_1, b_2, \dots, b_r, a_1 + c, a_2 + c, \dots, a_r + c].$$

This allows multigrades of higher order to be constructed from those of lower order. Because equal terms can be cancelled on both sides of a multigrade, the new one may have fewer than $2r$ terms on each side. For example, using $c = 5$ takes $[1, 6, 8] =^2 [2, 4, 9]$ to

$$[1, 6, 8, 7, 9, 14] =^3 [2, 4, 9, 6, 11, 13] \quad \text{or} \quad [1, 7, 8, 14] =^3 [2, 4, 11, 13].$$

Finding multigrades is often called the *Tarry-Escott Problem* or *Prouhet's Problem*. New multigrades continue to be searched for, and found.

CHAPTER 29

CARMICHAEL NUMBERS

Though it is sufficient that p be prime for $a^p \equiv a \pmod{p}$ to be true for all a, it is not necessary. There are composite numbers with that property.

Definition If n is composite and $a^n \equiv a \pmod{n}$ for all a, then n is a *Carmichael number.*

The first few Carmichael numbers are 561, 1105, 1729, 2465, and 2821.

The next theorem was proved by Korselt in 1899 but it was not until 1910 that the first Carmichael number was found.

Theorem *n is a Carmichael number if and only if n is not divisible by the square of any prime and $p \mid n$ implies $(p-1) \mid (n-1)$.*

Proof Suppose that n is a product of distinct primes, $n = p_1 p_2 \cdots p_r$, and $(p_i - 1) | (n-1)$ for each i. If we can show that $a^n - a \equiv 0 \pmod{p_i}$, i.e., that $p_i \mid (a^n - a)$ for each i, it follows that

$$p_1 p_2 \cdots p_r \mid (a^n - a),$$

i.e., that $a^n \equiv a \pmod{n}$.

If a is a multiple of p_i, then $a^n - a \equiv 0 \pmod{p_i}$. If a is not a multiple of p_i, then we know from Fermat's theorem that $a^{p_i - 1} \equiv 1 \pmod{p_i}$. Then, because $n - 1 = k(p_i - 1)$ for some k, we have

$$\begin{aligned} a^n - a &\equiv a^{(n-1)+1} - a \equiv a^{k(p_i-1)} \cdot a - a \\ &\equiv (a^{p_i-1})^k \cdot a - a \equiv 1^k \cdot a - a \\ &\equiv 0 \pmod{p_i}. \end{aligned}$$

So we have that $a^n \equiv a \pmod{n}$ for all a.

To prove the converse, suppose that $a^n \equiv a \pmod{n}$ for all a. If q is a prime and $q^2 \mid n$ then, because $q^n \equiv q \pmod{n}$, we know that $n \mid (q^n - q)$. Thus $q^2 \mid (q^n - q)$, which is impossible.

Suppose that $p \mid n$ and let g be a primitive root of p. Because $p \mid n$ and $n \mid (g^n - g)$ we know that $p \mid g(g^{n-1} - 1)$. Because g is a primitive root of p, $p \nmid g$, so $p \mid \left(g^{n-1} - 1\right)$. That is, $g^{n-1} \equiv 1 \pmod{p}$. We know that $n - 1$ must therefore be a multiple of the order of g, which is $p - 1$. That is, $(p-1) \mid (n-1)$.

Corollary *Carmichael numbers have at least three prime factors.*

Proof Suppose that $n = pq$ is a Carmichael number with p and q primes, $p < q$. Because $(q-1) | (n-1)$, $n = pq = k(q-1) + 1$. Modulo q, this gives $k \equiv 1 \pmod{q}$, or $k = 1 + jq$. Because $k > 1$, we have $j \geq 1$. Thus

$$n = (1 + jq)(q - 1) + 1 = q(jq - j + 1) = pq$$

so, cancelling q,

$$p = j(q - 1) + 1 > j(p - 1) + 1 \geq p - 1 + 1 = p,$$

which is impossible.

Corollary *If* $n = (6k+1)(12k+1)(18k+1)$ *where each factor is prime, then* n *is a Carmichael number.*

Proof n is not divisible by the square of any prime and $6k$, $12k$, and $18k$ all divide $n - 1 = 1296k^3 + 396k^2 + 36k$.

If $36k + 1$ is also prime, then $(6k + 1)(12k + 1)(18k + 1)(36k + 1)$ is a Carmichael number with four prime factors. It is likely that there are infinitely many Carmichael numbers of those two forms.

Be that as it may, it has been proved, though not until 1994, that there are infinitely many Carmichael numbers.

CHAPTER 30

Sophie Germain Primes

In the 1820s, Sophie Germain proved that if p is a prime such that $2p+1$ is also prime then $x^p + y^p = z^p$ had no solutions in positive integers relatively prime to p. Though her result has been superseded by Andrew Wiles' proof of Fermat's Last Theorem, her name has been given to such primes, which are called *Sophie Germain primes.* The first few are 2, 3, 5, 11, and 23. It is almost certainly true that there are infinitely many, though there is no proof of this. Exceedingly large ones continue to be discovered.

A sufficient condition for p to be a Sophie Germain prime is the

Theorem *If p is an odd prime and $q = 2p+1$ is such that $2^{q-1} \equiv 1 \pmod{q}$ and $3 \nmid q$ then q is a prime.*

Proof From one of our hypotheses we see that

$$4^p \equiv (2^2)^p \equiv 2^{2p} \equiv 2^{q-1} \equiv 1 \pmod{q}$$

so the order of 4 (mod q) is a divisor of p and thus 1 or p. By another of our hypotheses, $4^1 \equiv 1 \pmod{q}$ is impossible, so its order is p.

Suppose that q is composite. Then it has a smallest odd prime divisor r, which is not 3. Because $r \mid q$, the displayed congruence gives $4^p \equiv 1 \pmod{r}$. Because $r \neq 3$, this implies the order of 4 (mod r) cannot be 1 and hence must be p. From Fermat's Theorem, $4^{r-1} \equiv 1 \pmod{r}$. Thus $r-1$ is a multiple of p and so $r-1 \geq p$. So we have

$$r \geq p+1 > p + \frac{1}{2} = \frac{q}{2} > \sqrt{q},$$

the last inequality holding because $q \geq 7$. We have shown that the smallest prime divisor of q is greater than $\sqrt{q}$. If q is composite, its other prime divisors are also greater than $\sqrt{q}$ and the product of the prime factors of q will be greater than q. This is impossible, so q is prime.

Besides their use in Fermat's Last Theorem, Sophie Germain primes can be used to determine when some Mersenne numbers, that is integers of the form $2^p - 1$ with p prime, are composite.

Theorem *If p is a prime congruent to* 3 (mod 4) *then* $2p + 1$ *is prime if and only if* $(2p + 1) \mid (2^p - 1)$.

Proof Suppose that $q = 2p + 1$ is prime. Because $p = 4k + 3$, we have $q = 2(4k + 3) + 1 \equiv 7 \pmod 8$. We thus know that 2 is a quadratic residue of q so there exists n such that $n^2 \equiv 2 \pmod q$. Then Fermat's Theorem gives

$$2^p \equiv 2^{(q-1)/2} \equiv (n^2)^{(q-1)/2} \equiv n^{q-1} \equiv 1 \pmod q.$$

That is, $q \mid (2^p - 1)$, or $(2p + 1) \mid (2^p - 1)$.

To prove the converse, suppose that $(2p + 1) \mid (2^p - 1)$. If $2p + 1$ is composite it has a smallest prime factor r. Because $r \mid q$ the last displayed congruence gives $2^p \equiv 1 \pmod r$. The order of 2 $\pmod r$ is thus a divisor of p. Since the order is not 1, it is p. We know from Fermat's Theorem that $2^{r-1} \equiv 1 \pmod r$. Thus $r - 1$ is a multiple of p, so $r > p$. We know that $2p + 1 = rR$ for some $R \geq r$, so

$$2p + 1 = rR \geq r^2 > p^2,$$

which is impossible because $p \geq 3$. So $2p + 1$ is prime.

CHAPTER 31

The Group of Multiplicative Functions

A function f on the positive integers is multiplicative if $(m, n) = 1$ implies $f(mn) = f(m)f(n)$. We have seen some:

$$\sigma(n) = \sum_{d|n} d, \quad d(n) = \sum_{d|n} 1, \quad \phi(n) = n \prod_{p|n} \left(1 - \frac{1}{p}\right).$$

Here are some more:

$$i(n) = n, \quad s(n) = 1, \quad e(n) = \begin{cases} 1, & n = 1 \\ 0, & n \neq 1. \end{cases}$$

That they are multiplicative is not hard to verify.

We introduce an operation, the *Dirichlet convolution*, on multiplicative functions.

Definition If f and g are multiplicative functions, then

$$(f * g)(n) = \sum_{d|n} f(d)g(n/d).$$

For example,

$$(i * s)(n) = \sum_{d|n} i(d)s(n/d) = \sum_{d|n} i(d) = \sum_{d|n} d = \sigma(n).$$

Theorem *If f and g are multiplicative then $f * g$ is multiplicative.*

Proof If $(m, n) = 1$ and $d \mid mn$ then $d = d_1 d_2$ where $d_1 \mid m$ and $d_2 \mid n$. Then

$$\begin{aligned}(f * g)(mn) &= \sum_{d \mid mn} f(d)g(mn/d) \\ &= \sum_{d_1 d_2 \mid mn} f(d_1 d_2) g\big((m/d_1)(n/d_2)\big).\end{aligned}$$

Because $(d_1, d_2) = 1$ and $(m/d_1, n/d_2) = 1$ this becomes

$$\begin{aligned}\sum_{d_1 d_2 \mid mn} & f(d_1) f(d_2) g(m/d_1) g(n/d_2) \\ &= \sum_{d_1 \mid m} f(d_1) g(m/d_1) \sum_{d_2 \mid n} f(d_2) g(n/d_2) \\ &= (f * g)(m)(f * g)(n).\end{aligned}$$

Corollary *If f is multiplicative, then so is g, where $g(n) = \sum_{d \mid n} f(d)$.*

Proof $g = f * s$, and the convolution of two multiplicative functions is multiplicative.

The set of multiplicative functions is thus closed under $*$ and the operation can be shown to be commutative and associative. The identity element is e:

Theorem *If f is multiplicative, then $f * e = f$.*

Proof $(f * e)(n) = \sum_{d \mid n} f(d) e(n/d) = f(n)$.

There are inverses:

Theorem *If f is multiplicative and not identically* 0 *then there is g such that $f * g = e$.*

Proof First we note that $f(1) = 1$. Because f is not identically 0, there is an integer k such that $f(k) \neq 0$. But $f(k) = f(k \cdot 1) = f(k) f(1)$ and cancelling $f(k)$ gives the result.

We want $(f * g)(1) = e(1) = 1$ so $f(1)g(1) = 1$ and $g(1) = 1$. We now proceed by mathematical induction. We have the value of g at 1.

Suppose that $n > 1$ and that we know the value of $g(m)$ for all m with $m < n$. We want to have

$$(f * g)(n) = \sum_{d|n} f(d)g(n/d) = e(n) = 0.$$

This is true if

$$f(1)g(n) + \sum_{d|n, d>1} f(d)g(n/d) = 0.$$

By the induction assumption, we know all the values of g in the sum so we can solve the equation for $g(n)$. Thus all the values of g are determined.

So, multiplicative functions (excluding $f(n) = 0$ for all n) under $*$ form a commutative group. Its elements satisfy such relations as $s * s = d$ and $\phi * s = i$. An important member of the group is the inverse of s:

Definition The *Möbius function* μ is the inverse of s.

Theorem $\mu(1) = 1$, $\mu(n) = 0$ *if* n *is divisible by a square, and, if* n *is a product of* k *distinct primes,* $\mu(n) = (-1)^k$.

Proof Because μ is multiplicative its values are determined by its values at prime powers. If p is prime,

$$e(p) = \sum_{d|p} s(d)\mu(p/d)$$

or

$$0 = s(1)\mu(p) + s(p)\mu(1) = \mu(p) + \mu(1)$$

whence $\mu(p) = -1$. For p^2,

$$\begin{aligned} 0 &= s(1)\mu(p^2) + s(p)\mu(p) + s(p^2)\mu(1) \\ &= \mu(p^2) - 1 + 1 \end{aligned}$$

so $\mu(p^2) = 0$. In general, $0 = \sum_{i=1}^{k} \mu(p^i)$ and proceeding by mathematical induction gives $\mu(p^k) = 0$ for $k \geq 2$. Thus, if $n = p_1^{e_1} p_2^{e_2} \cdots p_r^{e_r}$, if any of the exponents is 2 or more then $\mu(n)$, being the product of the $\mu(p_i^{e_i})$ terms, will be zero. If n is a product of k distinct primes, then $\mu(n)$ is a product of k terms, $\mu(p_i) = -1$, and thus $(-1)^k$.

The Möbius function gives the

Theorem (the Möbius inversion formula) *If g is multiplicative and*

$$f(n) = \sum_{d|n} g(d)$$

then

$$g(n) = \sum_{d|n} f(d)\mu(n/d).$$

Proof The hypothesis is that $f = g * s$, so

$$f * \mu = g * (s * \mu) = g * e = g$$

which is the conclusion.

CHAPTER 32

BOUNDS FOR $\pi(x)$

Let $\pi(x)$ denote the number of primes less than or equal to x. The Prime Number Theorem, that $\pi(x)$ is asymptotic to $x/\ln x$ in the sense that

$$\lim_{x\to\infty} \frac{\pi(x)}{x/\ln x} = 1$$

(actually, $\int_2^x dt/\ln t$ is better than $x/\ln x$, as was noticed by Gauss in 1792), was not proved until 1896. More details can be found in Chapter 35. The proof of the theorem, as might be expected of something that eluded Gauss, is difficult. Here we will get a weaker result, namely the

Theorem *For $x \geq 2$,*

$$\frac{1}{4}\ln 2 < \frac{\pi(x)}{x/\ln x} < 32\ln 2.$$

We will be following work by Tchebyshev, who established the inequality with upper and lower bounds considerably closer to 1.

The key to the proof, here and in the proofs of the Prime Number Theorem, is relating $\pi(x)$ to something that can be dealt with independently from the primes, in this case the binomial coefficient $\binom{2n}{n}$. Examining the prime-power decomposition of $\binom{2n}{n}$ for small values of n can lead to many conjectures, including

All the primes between n and $2n$ appear with exponent 1.
None of the primes between $2n/3$ and n appear at all.
$\binom{2n}{n}$ is divisible by $n+1$.
No prime power has value greater than $2n$.

All of these are true, and can be observed in

$$\binom{38}{19} = 2^3 \cdot 3 \cdot 5^2 \cdot 7 \cdot 11 \cdot 23 \cdot 29 \cdot 31 \cdot 37.$$

The last property is not as easy to verify as the first three, but it is the one that we need. Because

$$\binom{2n}{n} = \frac{(2n)!}{n!n!}$$

we start by looking at the prime-power decomposition of $n!$.

Lemma *The exponent of p in the prime-power decomposition of $n!$ is*

$$\left[\frac{n}{p}\right] + \left[\frac{n}{p^2}\right] + \left[\frac{n}{p^3}\right] + \cdots.$$

Proof Each multiple of p that is less than or equal to n contributes one factor of p to the prime-power decomposition of $n!$. There are $[n/p]$ such multiples because

$$\frac{n}{p} = \left[\frac{n}{p}\right] + \frac{r}{p}, \qquad 0 \leq r < p.$$

Each multiple of p^2 that is less than or equal to n contributes another factor of p, and there are $[n/p^2]$ of those. And so on. The sum will terminate because sooner or later n/p^k will be less than 1.

Corollary *The exponent of p in the prime-power decomposition of $\binom{2n}{n}$ is*

$$\left(\left[\frac{2n}{p}\right] - 2\left[\frac{n}{p}\right]\right) + \left(\left[\frac{2n}{p^2}\right] - 2\left[\frac{n}{p^2}\right]\right) + \cdots.$$

Proof We apply the lemma to

$$\binom{2n}{n} = \frac{(2n)!}{(n!)^2}.$$

The power of p in its prime-power decomposition is the power in $(2n)!$ divided by the power in $(n!)^2$. The exponent of the power in the numerator is $[2n/p] + [2n/p^2] + \cdots$ and the exponent of the power in the denominator is $2([n/p] + [n/p^2] + \cdots)$, which gives the result.

Lemma *Each prime power in the prime-power decomposition of $\binom{2n}{n}$ is less than or equal to $2n$.*

Proof Suppose that p^r appears in the prime-power decomposition of $\binom{2n}{n}$ and that $p^r > 2n$. Then

$$[2n/p^r] = \left[2n/p^{r+1}\right] = \cdots = 0$$

and

$$[n/p^r] = \left[n/p^{r+1}\right] = \cdots = 0.$$

So, the sum in the Corollary terminates after r terms. Each term in the sum has the form $[2x] - 2[x]$. Because $[2x] \leq 2x$ and $[x] > x - 1$, we know that

$$[2x] - 2[x] < 2x - 2(x - 1) = 2.$$

So, each term has value at most 1 and there are $r - 1$ terms. The sum is then less than or equal to $r - 1$. But the sum is the exponent of p in the prime-power decomposition, which is r. We have shown that $r \leq r - 1$. Because this is impossible, our assumption that $p^r > 2n$ is incorrect, so $p^r \leq 2n$.

We next need bounds on $\binom{2n}{n}$.

Lemma *For* $n = 1, 2, \ldots,$

$$2^n \leq \binom{2n}{n} \leq 2^{2n}.$$

Proof We use mathematical induction. The lemma is true for $n = 1$. Suppose that it is true for $n = k$. Then

$$\begin{aligned}\binom{2(k+1)}{k+1} &= \frac{(2k+2)!}{((k+1)!)^2} \\ &= \frac{(2k+2)(2k+1)(2k)!}{(k+1)^2(k!)^2} \\ &= \frac{2(2k+1)}{(k+1)}\binom{2k}{k}.\end{aligned}$$

Because $2 < 2(2k+1)/(k+1) < 4$ we have that it is true for $n = k + 1$.

Lemma *For* $n \geq 2$,

$$\pi(2n) - \pi(n) \leq \frac{2n \ln 2}{\ln n}.$$

Proof Because

$$\binom{2n}{n} = \frac{(2n)(2n-1)\cdots(n+1)}{n(n-1)\cdots 1}$$

no factor in the numerator that is a prime can be cancelled by anything in the denominator. Thus, using the last lemma,

$$2^{2n} \geq \binom{2n}{n} \geq \prod_{n<p\leq 2n} p \geq \prod_{n<p\leq 2n} n = n^{\pi(2n)-\pi(n)}.$$

Taking logarithms gives the result.

Lemma *For $n \geq 2$,*

$$\pi(2n) \geq (n \ln 2)/\ln 2n.$$

Proof We know that each prime power in the prime-power decomposition of $\binom{2n}{n}$ is at most $2n$, and there are at most $\pi(2n)$ of them. Thus

$$2^n \leq \binom{2n}{n} \leq (2n)^{\pi(2n)}.$$

Taking logarithms gives the result.

Lemma *For $r \geq 1$,*

$$\pi(2^{2r}) < \frac{2^{2r+2}}{r}.$$

Proof We will use mathematical induction. The lemma is true for $r = 1$ because $\pi(4) < 16$. Suppose that it is true for $r = k$. Then, applying the lemma before last,

$$\begin{aligned}\pi(2^{2k+2}) &= \left(\pi(2^{2k+2}) - \pi(2^{2k+1})\right) + \left(\pi(2^{2k+1}) - \pi(2^{2k})\right) + \pi(2^{2k}) \\ &\leq \frac{2 \cdot 2^{2k+1} \ln 2}{\ln 2^{2k+1}} + \frac{2 \cdot 2^{2k} \ln 2}{\ln 2^{2k}} + \pi(2^{2k}).\end{aligned}$$

By applying the induction assumption and doing some arithmetic, we see that this comes out to be less than $(7 \cdot 2^{2k})/k$. Because $7/k$ is less than $16/(k+1)$ for $k \geq 1$, we have the result.

Theorem *For $x \geq 2$,*

$$\frac{1}{4} \ln 2 \frac{x}{\ln x} \leq \pi(x) \leq 32 \ln 2 \frac{x}{\ln x}.$$

Proof For the right-hand inequality, choose r such that $2^{2r-2} \leq x < 2^{2r}$ and apply the last lemma:

$$\frac{\pi(x)}{x} \leq \frac{\pi(2^{2r})}{2^{2r-2}} < \frac{2^{2r+2}}{2^{2r-2}r} = \frac{16}{r}.$$

But $\ln x < (2r)\ln 2$, so $r > (\ln x)/(2\ln 2)$ and it follows that $\pi(x)/x < (32\ln 2)/\ln x$.

For the left-hand inequality, let n be so that $2n \leq x < 2n+2$ and apply the next-to-last lemma:

$$\pi(x) \geq \pi(2n) \geq \frac{n\ln 2}{\ln 2n} \geq \frac{n\ln 2}{\ln x} \geq \frac{2n+2}{4}\frac{\ln 2}{\ln x} > \frac{x}{4}\frac{\ln 2}{\ln x}.$$

No proof of the Prime Number Theorem (there are several), or of anything approaching it, is simple.

CHAPTER 33

The Sum of the Reciprocals of the Primes

We know that $\sum_{n=1}^{\infty} 1/n$ diverges so, because the primes are so much rarer than integers—the gaps between successive primes can be arbitrarily large—we might think that $\sum_{p \text{ prime}} 1/p$ has a chance of converging.

The Prime Number Theorem indicates that this would be mistaken.

Theorem $\displaystyle\sum_p \frac{1}{p}$ *diverges.*

Plausibility Argument The Prime Number Theorem says that

$$\pi(x) \sim \frac{x}{\ln x}.$$

Because

$$\ln \pi(x) \sim \ln \frac{x}{\ln x} = \ln x - \ln(\ln x)$$

we have

$$\frac{\ln \pi(x)}{\ln x} \sim 1 - \frac{\ln(\ln x)}{\ln x},$$

which goes to 1 as $x \to \infty$. Thus $\ln x \sim \ln \pi(x)$ and so $\pi(x) \sim \frac{x}{\ln \pi(x)}$. That is,

$$x \sim \pi(x) \ln \pi(x).$$

Let $x = p_n$, the nth prime. Then $\pi(x) = \pi(p_n) = n$ so this becomes

$$p_n \sim n \ln n$$

and so

$$\sum_{n=1}^{N} \frac{1}{p_n} \sim \sum_{n=1}^{N} \frac{1}{n \ln n} \sim \int_1^N \frac{dx}{x \ln x} \sim \ln(\ln N).$$

As $N \to \infty$, $\ln(\ln N) \to \infty$ (slowly) so the sum of the reciprocals of the primes diverges.

We will make this a bit more rigorous by using an important identity of Euler:

Theorem $\displaystyle\sum_{n=1}^{\infty} \frac{1}{n} = \prod_{p \text{ prime}} \frac{1}{1-p^{-1}}.$

Proof It might be objected that, because the series on the left diverges, the equation makes no sense. It did to Euler, however, and the theorem can be brought into compliance with present-day standards of rigor by restricting the sum and product to a finite number of terms. The details needed to do that take up space and do not enhance the idea, and so are omitted.

We know that

$$\frac{1}{1-p^{-1}} = 1 + p^{-1} + p^{-2} + \cdots,$$

so

$$\prod_p \frac{1}{1-p^{-1}} = \prod_p \left(1 + \frac{1}{p} + \frac{1}{p^2} + \frac{1}{p^3} + \cdots\right).$$

When we multiply the right-hand side out, we get a collection of terms obtained by taking one factor from each parenthesis, for $p = 2, 3, 5, \ldots$:

$$\frac{1}{2^{e_1} \cdot 3^{e_2} \cdot 5^{e_3} \ldots}.$$

The exponents are greater than or equal to zero. There are no terms with an infinite number of non-zero exponents because they have value 0 (this can be made rigorous) so we have precisely the reciprocals of the prime-power decompositions of each positive integer. Thus

$$\prod_p \left(1 + \frac{1}{p} + \frac{1}{p^2} + \frac{1}{p^3} + \cdots\right) = \sum_{n=1}^{\infty} \frac{1}{n}.$$

We can use this to get our result.

Proof of divergence The sums and products are infinite, but the proof could be modified so as to consider only finite sums and products.

Taking logarithms in Euler's identity, we have

$$\begin{aligned}
\ln\left(\sum_{n=1}^{\infty}\frac{1}{n}\right) &= \ln\left(\prod_{p}\frac{1}{1-p^{-1}}\right) \\
&= \sum_{p}\ln\frac{1}{1-p^{-1}} = \sum_{p} -\ln(1-p^{-1}) \\
&= \sum_{p}\left(\frac{1}{p}+\frac{1}{2p^2}+\frac{1}{3p^3}+\cdots\right) \\
&= \sum_{p}\frac{1}{p}+\sum_{p}\frac{1}{p^2}\left(\frac{1}{2}+\frac{1}{3p}+\frac{1}{4p^2}+\cdots\right) \\
&< \sum_{p}\frac{1}{p}+\sum_{p}\frac{1}{p^2}\left(1+\frac{1}{p}+\frac{1}{p^2}\cdots\right) \\
&= \sum_{p}\frac{1}{p}+\sum_{p}\frac{1}{p^2}\left(\frac{1}{1-p^{-1}}\right) \\
&= \sum_{p}\frac{1}{p}+\sum_{p}\frac{1}{p(p-1)} \\
&< \sum_{p}\frac{1}{p}+\sum_{n=2}^{\infty}\frac{1}{n(n-1)} \\
&= \sum_{p}\frac{1}{p}+1.
\end{aligned}$$

Because the left-hand side is not finite,

$$\sum_{p}\frac{1}{p}$$

cannot be finite.

CHAPTER 34

THE RIEMANN HYPOTHESIS

Definition If the real part of the complex variable $s = \varsigma + it$ is greater than 1, the *Riemann zeta function*, $\varsigma(s)$, is

$$\sum_{n=1}^{\infty} \frac{1}{n^s}.$$

When $\Re(s) > 1$ the series converges. Its connection with the primes comes from Euler's

Theorem *If* $\Re(s) > 1$,

$$\begin{aligned} \varsigma(s) &= \sum_{n=1}^{\infty} \frac{1}{n^s} \\ &= \prod_{p} \frac{1}{1 - p^{-s}}. \end{aligned}$$

Semi-proof The proof uses the same idea as was used in the semi-proof of the corresponding statement in Chapter 33 where 1 appeared instead of s. Questions of convergence must be dealt with, which we do not do here.

Euler proved the

Theorem *If* $\Re(s) > 0$,

$$\varsigma(s) = \frac{1}{1 - 2^{1-s}} \sum_{n=1}^{\infty} \frac{(-1)^{n-1}}{n^s}.$$

Partial Proof If $\Re(s) > 1$,

$$\begin{aligned}\varsigma(s) - \sum_{n=1}^{\infty} \frac{(-1)^{n-1}}{n^s} &= \sum_{n=1}^{\infty} \frac{1-(-1)^{n-1}}{n^s} \\ &= \sum_{n \text{ even}} \frac{2}{n^s} \\ &= \sum_{m=1}^{\infty} \frac{2}{(2m)^s} = 2^{1-s}\varsigma(s).\end{aligned}$$

When this is extended to $\Re(s) > 0$ we have the zeta function defined in the right-hand half plane except at $s = 1$, where it has a simple pole.

Riemann extended the domain of the zeta function to the left-hand half plane with the reflection equation

$$\varsigma(s) = 2^s \pi^{s-1} \sin(\pi s/2)\Gamma(1-s)\varsigma(1-s).$$

Here $\Gamma(s) = \int_0^{\infty} x^{s-1}e^{-x}dx$ is the gamma function. A symmetric version of the result is

$$\pi^{-s/2}\Gamma(s/2)\varsigma(s) = \pi^{-(1-s)/2}\Gamma\big((1-s)/2\big)\varsigma(1-s).$$

A proof of the Prime Number Theorem involves showing that

$$\lim_{x\to\infty} \frac{1}{x} \sum_{\rho} \frac{x^{\rho}}{\rho} = 0$$

where the sum is taken over all the zeros of the zeta function in the strip $0 < \varsigma < 1$. To show this, knowledge of the location of the zeros is necessary.

Though not needed for the proof of the PNT, Riemann conjectured

Conjecture (the Riemann Hypothesis) *The zeros of the zeta function in the strip $0 < \sigma < 1$ all lie on the line $\sigma = 1/2$.*

If the hypothesis is true, which hardly anyone doubts, many results will follow. The first few billion zeros of the zeta function in the strip have been found to lie on the line. Because there is a $1,000,000 prize for settling the conjecture many proofs have been put forward recently, some by reputable mathematicians, but all have been found wanting.

CHAPTER 35

THE PRIME NUMBER THEOREM

The Prime Number Theorem is that $\pi(x)$, the number of primes less than or equal to x, is asymptotically equal to $x/\ln x$ (so that

$$\lim_{x\to\infty} \frac{\pi(x)}{x/\ln x} = 1)$$

or, as Gauss observed, to $\int_2^\infty \frac{dt}{\ln t}$. Riemann and Tchebyshev laid the groundwork for the proof, which was finally achieved in 1896 by Hadamard and de la Vallée Poussin. Here is an extremely brief sketch of the ideas in the proof. Pages of details have been omitted.

The *Mangoldt function* is defined by

$$\Lambda(n) = \begin{cases} \ln p, & n = p^k,\ k \geq 1 \\ 0, & \text{otherwise.} \end{cases}$$

It has a connection with the Riemann zeta function, which does not involve the primes in its definition:

$$-\frac{\zeta'(s)}{\zeta(s)} = \sum_{n=1}^{\infty} \frac{\Lambda(n)}{n^s}.$$

Let $\psi(x) = \sum_{n\leq x} \Lambda(n)$. It can be shown that $\pi(x) \sim x/\ln x$ if and only if $\psi(x) \sim x$. It can also be shown that

$$\psi(x) = -\frac{1}{2\pi i} \int_{c-i\infty}^{c+i\infty} \frac{\zeta'(s)}{\zeta(s)} x^s \frac{ds}{s}.$$

By moving c to the left as far as possible, we can get (essentially)

$$\psi(x) = x - \sum_{\rho} \frac{x^{\rho}}{\rho} - \frac{\varsigma'(0)}{\varsigma(0)} - \frac{1}{2}\ln(1 - x^{-2}),$$

where the sum is taken over all the zeros of the zeta function in the strip $0 < \varsigma < 1$. From this $\psi(x) \sim x$ can be shown to follow.

Proofs not using the zeta function or complex variables, and hence called elementary, were given in 1948 by Erdős and Selberg.

CHAPTER 36

THE *abc* CONJECTURE

If a and b are relatively prime and $c = a + b$, it is usually the case that c is smaller than the product of the primes that divide a, b, and c, as $8+25 = 33$ and $33 < 2 \cdot 5 \cdot 3 \cdot 11$. This does not always happen, however, as is shown by $5 + 27 = 32$ and $32 > 5 \cdot 3 \cdot 2$. The exceptions are sufficiently rare that in 1985 Masser and Oesterlé made the *abc conjecture*:

Conjecture *Given $\varepsilon > 0$ there is a number k, which depends on ε, such that*

$$c < k\Bigl(\prod_{p|abc} p\Bigr)^{1+\varepsilon}$$

is true for all a, b, c with $a + b = c$ and $(a, b) = 1$.

The conjecture is very powerful. If true, it would settle a large number of open problems.

To illustrate its power, here is how a weaker version would prove Fermat's Last Theorem.

Theorem *If a and b are relatively prime, $a+b = c$, and $c < \left(\prod_{p|abc} p\right)^2$ for all a, b, and c, then $x^n + y^n = z^n$ has no solution in positive integers for $n \geq 3$.*

Proof Let $a = x^n$, $b = y^n$, and $c = z^n$. Then

$$z^n < \Bigl(\prod_{p|xyz} p\Bigr)^2 < (xyz)^2 < z^6.$$

So $n < 6$. But it has been known for a long time that $x^n + y^n = z^n$ has no positive integer solutions for $n = 3$, 4, or 5, thus proving the result.

It is not likely that the *abc* conjecture will be verified any time soon, but one never knows.

Similar is *Schinzel's Hypothesis H*:

Conjecture *If $f_1(x), f_2(x), \dots, f_k(x)$ are irreducible polynomials with integer coefficients and no integer $n > 1$ divides $f_1(x)f_2(x)\cdots f_k(x)$ for all integers x, then there are infinitely many x such that the polynomials are simultaneously prime.*

If true, the hypothesis would also answer many open questions. In particular, applying it to x and $x+2$ would show that there are infinitely many twin primes. The hypothesis was formulated in 1958 and there are no signs of a proof.

CHAPTER 37

Factorization and Testing for Primes

To see if an integer N is prime, it is always possible to try divisors $2, 3, 5, \ldots$ up to the last prime less than or equal to $\sqrt{N}$. If none of them divides N, then N is prime.

The test can take as many as $\sqrt{N}$ steps, which is large if N is large. A test devised by Lucas is the basis for some prime-testing algorithms that are more efficient.

Theorem *If* $a^{n-1} \equiv 1 \pmod{n}$ *and if, for each prime* p *that divides* $n-1$, $a^{(n-1)/p} \not\equiv 1 \pmod{n}$, *then* n *is prime.*

Proof Let t be the order of $a \pmod{n}$, so that $a^t \equiv 1 \pmod{n}$ and t is the smallest positive integer with that property. Then we know that $n-1$ is a multiple of t, so $n-1 = tr$ for some r. Suppose that $r > 1$. Then there is a prime p that divides r, and hence divides $n-1$ as well. But then

$$\begin{aligned} a^{(n-1)/p} &\equiv a^{tr/p} \\ &\equiv (a^t)^{r/p} \\ &\equiv 1^{r/p} \\ &\equiv 1 \pmod{n}, \end{aligned}$$

which is impossible. Thus our assumption that $r > 1$ was incorrect and $r = 1$. So t, the order of a, is $n-1$. Because $a^{\phi(n)} \equiv 1 \pmod{n}$, we know that $\phi(n)$ is a multiple of $n-1$. Because $\phi(n) \leq n-1$, the multiplier must be 1 and $\phi(n) = n-1$. This implies that n is prime.

The largest known primes are Mersenne numbers because there is the Lucas-Lehmer test, relatively easy to apply, for numbers of the form $2^p - 1$:

Theorem *Suppose that* $N = 2^p - 1$ *for a prime* p. *Let* $s_0 = 4$ *and* $s_n \equiv s_{n-1}^2 - 2 \pmod{N}$, $n = 1, 2, \ldots$. *Then* N *is prime if and only if* $s_{p-2} \equiv 0 \pmod{N}$.

The proof is fairly complicated.

Factoring large numbers is more difficult than testing them for primality. One idea is to look at $a^2 - N$ for integers a starting with $\left[\sqrt{N}\right]$. If the difference is ever a square, we have $a^2 - N = b^2$ and a factorization $N = (a+b)(a-b)$. If $a - b = 1$, we have to look further if we want to get anything useful.

A similar idea is to search for a and b so that $a^2 \equiv b^2 \pmod{N}$. For such integers, if $p \mid N$ then $p \mid \left(a^2 - b^2\right)$ and hence $p \mid (a+b)$ or $p \mid (a-b)$. If either of the greatest common divisors $(a+b, N)$ or $(a-b, N)$, easily calculated using the Euclidean algorithm, is different from 1 and N, a factor of N has been discovered.

Pollard's factorization algorithm uses the greatest common divisor idea. To try to discover a factor of N take a function, say $f(x) = x^2 + 1$ (other choices are possible), some initial values, say $x_0 = 2$, $y_0 = 2$, and construct two sequences,

$$\begin{aligned} x_k &\equiv f(x_{k-1}) \equiv x_{k-1}^2 + 1 \pmod{N} \\ y_k &\equiv f\left(f(y_{k-1})\right) \equiv f(y_{k-1}^2 + 1) \\ &\equiv y_{k-1}^4 + 2y_{k-1}^2 + 2 \pmod{N}, \end{aligned}$$

$k = 1, 2, \ldots$. If $x_k = y_k$, we have been unlucky and have to quit and try again, with a different function or different initial values. If $x_k \neq y_k$, look at $(|x_k - y_k|, N)$. If its value is 1, proceed to the next k. If it is anything else, we have found a factor of N.

All these methods depend on computers to do the calculations.

Long before computers, Fermat found a method of factorization: if an integer is a sum of two squares in two different ways, then it can be factored.

Theorem *If* N *has two different representations as a sum of two squares,* $N = r^2 + s^2 = t^2 + u^2$ *(so* $s \neq t$ *and* $s \neq u$*), then* N *is composite.*

Proof We have $(r+t)(r-t) = (u-s)(u+s)$. Let $d = (r-t, u-s)$, so

$$r - t = dj, \quad u - s = dk$$

for relatively prime integers j, k. Then $(dj)(r+t) = (dk)(u+s)$ or

$$j(r+t) = k(u+s).$$

Because $(j, k) = 1$, this says that $k \mid (r+t)$ so for some l

$$r + t = kl$$

and so $j(kl) = k(u+s)$, or $u + s = jl$. Thus

$$2r = dj + kl, \quad 2s = jldk$$

and

$$\begin{aligned} 4N &= 4r^2 + 4s^2 \\ &= (d^2j^2 + 2jkdl + k^2l^2) + (j^2l^2 - 2jkdl + d^2k^2) \\ &= (d^2 + l^2)(j^2 + k^2). \end{aligned}$$

If one of the pairs (d, l), (j, k) consists of even integers, dividing by 4 gives two factors of N. One will: if we name s and u so they have the same parity, r and t will have the same property so that d will be even. Then either l or both of j and k are even.

For example, let us factor $50 = 7^2 + 1^2 = 5^2 + 5^2$. Here $r = 7$, $s = 1$, $t = 5$, and $u = 5$. We then get that $d = (2, 4) = 2$, $j = 1$, $k = 2$, and $l = 6$, leading to

$$4 \cdot 50 = (2^2 + 6^2)(1^2 + 2^2) = 40 \cdot 5$$

and hence to $50 = 10 \cdot 5$. We knew that already, but larger integers can lead to less obvious factorizations.

Because of the importance of primes and factorization in methods of encrypting data, much effort has gone into finding methods to test for primality or to find factors, many of which exist that are not mentioned here.

CHAPTER 38

Algebraic and Transcendental Numbers

Though almost all numbers are transcendental, transcendental numbers can be hard to identify.

Definition A number is *algebraic* if it is the root of a polynomial equation with integer coefficients.

So $\sqrt{2}$ and i are algebraic, being roots of $x^2 - 2 = 0$ and $x^2 + 1 = 0$, as are all the roots of $x^5 + x + 1 = 0$.

The roots of a polynomial equation with rational coefficients are also algebraic, because multiplication by a common denominator can make the coefficients integers. It is a fact (that we will not prove) that if the coefficients are algebraic numbers, then so are the roots.

Theorem *There are only countably many algebraic numbers.*

Proof For each integer n, there are only countably many polynomials of degree n with integer coefficients, so, because each has only finitely many roots, we get only countably many algebraic numbers from polynomials of degree n. Since the set of degrees, $\{1, 2, \dots\}$, is countable, so is the totality of algebraic numbers.

Definition A real number that is not algebraic is *transcendental.*

There are uncountably many real numbers. Because there are only countably many algebraic numbers (which include the integers and rational numbers), almost all real numbers are transcendental.

The first examples of transcendental numbers were found by Liouville in 1844. They come from a corollary of his theorem that rational numbers cannot get too close to irrational algebraic numbers, in the following sense:

Theorem *If a is a real irrational algebraic number, the root of a polynomial equation of degree n, then there is a positive constant k, which depends on a, such that*

$$\left|a - \frac{r}{s}\right| > \frac{k}{s^n}$$

for rational numbers r/s with $s > 0$.

Sketch of Proof Suppose that $f(a) = 0$ where f is a polynomial of degree n with integer coefficients. For integers r and s such that $r/s, s > 0$, is close to a, $|f(r/s)|$ is a positive rational number with denominator s^n and thus has a value at least as large as $1/s^n$. When r/s is near a we know that

$$f(a) - f(r/s) \approx (a - r/s) \cdot f'(a).$$

Thus

$$\left|a - \frac{r}{s}\right| \approx \left|\frac{0 - f(r/s)}{f'(a)}\right| = k\,|f(r/s)| \geq k \cdot \frac{1}{s^n}.$$

A real irrational number that can be well approximated by rational numbers is thus transcendental.

Corollary $a = \sum_{j=1}^{\infty} 10^{-j!}$ *is transcendental.*

Proof The series converges (very quickly!). Suppose that a is algebraic of degree n. Then there exists k such that $\left|a - \frac{r}{s}\right| > \frac{k}{s^n}$ for all r and s. Choose c large enough so that $1/2^c < k$. Choose $j > c + n$. Then $\sum_{i=1}^{j} 10^{-i!}$ is a rational number r/s, with $s = 10^{j!}$. Thus

$$\begin{aligned}\left|a - \frac{r}{s}\right| &= \sum_{i=j+1}^{\infty} \frac{1}{10^{i!}} < \frac{1}{10^{(j+1)!}}\left(1 + \frac{1}{10} + \frac{1}{10^2} + \cdots\right) \\ &= \frac{10}{9}\frac{1}{s^{j+1}} < \frac{1}{s^j} < \frac{1}{s^{c+n}} < \frac{1}{2^c}\frac{1}{s^n} < \frac{k}{s^n},\end{aligned}$$

which is a contradiction.

Many other numbers can be defined similarly and shown to be transcendental.

Hermite showed in 1873 that e is transcendental. In 1882 Lindemann showed that π is transcendental, thus putting to rest the problem of squaring the circle with straightedge and compass because every number that can be constructed with those tools is algebraic.

A general result was proved by Gelfond and Schneider in 1934:

Theorem *If a is algebraic but not 0 or 1, and b is algebraic but not a real rational number, then a^b is transcendental.*

The proof, as might be expected, is not simple. The theorem shows that $2^{\sqrt{2}}$ is transcendental, as is e^{π} because it is one of the values of $(-1)^{-i} = (e^{\pi i})^{-i}$.

The status of most numbers remains unknown. A simple result is the

Theorem *If a and b are transcendental, at least one of $a + b$ and ab is also.*

Proof If $a + b$ and ab are algebraic, then the roots of

$$0 = x^2 - (a + b)x + ab = (x - a)(x - b),$$

namely a and b, are also algebraic, but this is a contradiction.

Because the odds are ∞ to 1 that $a + b$ and ab are both transcendental, the theorem is not all that helpful, but such is the state of knowledge of transcendental numbers.

CHAPTER 39

UNSOLVED PROBLEMS

Number theory abounds with problems that are easy to state but hard to solve. The Riemann Hypothesis and the *abc* Conjecture have already been mentioned. There follow some more.

The *Goldbach Conjecture*, made by Goldbach in 1742, is that every even number greater than 2 is a sum of two primes. This was not a conjecture made for conjecture's sake: Goldbach was trying to help Euler find a proof of the theorem that every integer is a sum of four squares. The conjecture, though undoubtedly true, has turned out to be harder than the four-squares theorem. In 1973 it was shown that every sufficiently large even integer is a sum of a prime and a number that has at most two prime factors.

The primes thin out as we proceed through the integers, but as far out as anyone has looked there are primes p such that $p + 2$ is also prime. The *twin prime conjecture*, also undoubtedly true, is that there are infinitely many such pairs. A similar conjecture is that there are infinitely many prime quadruples—p, $p + 2$, $p + 6$, $p + 8$ all prime.

The form of even perfect numbers was determined by Euler. It is not known if any odd perfect numbers exist. If one does, it must be very large, and the conjecture is that there is none. My opinion is that there is one—infinitely many, in fact—but it is too large for us ever to find.

It is surely true that there are infinitely many Mersenne primes, those of the form $2^p - 1$ with p prime, but no proof is in sight. There is also a conjecture that $2^p - 1$, prime or not, is always squarefree. Counterexamples may be too large to find.

Take an integer. If it is even, divide it by two. If it is odd, triple it and add one. Repeat. Applied to 14, we get 7, 22, 11, 34, 17, 52, 26, 13, 40, 20, 10, 5, 16, 8, 4, 2, 1, 4, 2, 1,.... The *Collatz problem* is to determine if this process always leads to 1.

Let

$$\gamma = \lim_{n\to\infty}\left(\sum_{k=1}^{n}\frac{1}{k} - \ln n\right)$$

be Euler's constant, 0.57721566.... Is it irrational? Probably, but no one knows for sure. Here is the first sentence of the proof that it is: "Suppose that $\gamma = r/s$ for some integers r and s." Take it from there.

Is there a prime between n^2 and $(n+1)^2$ for every positive integer n?

Fermat numbers $2^{2^n} + 1$ are prime for $n = 0, 1, 2, 3$, and 4 but have turned out to be composite for all larger values of n for which their nature has been determined. The conjecture that all of them are composite for $n \geq 4$ is tempting to make, has been made, and is unlikely to be settled any time soon.

Are there infinitely many n such that $\phi(n) = \phi(n+1)$, where ϕ denotes Euler's ϕ-function?

The list could be extended, in fact to almost any length. The latest edition of Richard Guy's *Unsolved Problems in Number Theory* runs to more than 400 pages.

Unsolved problems sometimes get solved, the example of Fermat's Last Theorem being the most obvious. Recent examples are the proof of Catalan's conjecture that 8 and 9 are the only consecutive perfect powers by Mihăilescu and the proof by Tao and Green that there are arbitrarily long arithmetic progressions of primes.

Index

About the Author

Underwood Dudley earned his B.S. and M.S. degrees from the Carnegie Institute of Technology and his doctorate (in number theory) from the University of Michigan. He taught briefly at the Ohio State University and then at DePauw University from 1967–2004. Woody has written six books and many papers, reviews, and commentaries. He has served in many editing positions, including editor of *The Pi Mu Epsilon Journal*, 1993–96 and *The College Mathematics Journal*, 1999–2003. He is widely known and admired for his speaking ability—especially his ability to find humor in mathematics. He was the PME J. Sutherland Frame lecturer in 1992 and the MAA Pólya lecturer in 1995–96. Woody's contributions to mathematics have earned him many awards, including the Trevor Evans award, from the MAA in 1996, the Distinguished Service Award, from the Indiana Section of the MAA in 2000, and the Meritorious Service Award, from the MAA in 2004.